MANAGEMENT INFORMATION SYSTEMS

WHARTON EXECUTIVE LIBRARY

The Wharton School, University of Pennsylvania

F. Gerard Adams
The Business Forecasting Revolution: Nation–Industry–Firm

James C. Emery
Management Information Systems: The Critical Strategic Resource

Leonard M. Lodish
The Advertising and Promotion Challenge: Vaguely Right or Precisely Wrong?

David Solomons
Making Accounting Policy: The Quest for Credibility in Financial Reporting

MANAGEMENT INFORMATION SYSTEMS

The Critical Strategic Resource

James C. Emery

New York • *Oxford*
OXFORD UNIVERSITY PRESS
1987

Oxford University Press

Oxford New York Toronto
Delhi Bombay Calcutta Madras Karachi
Petaling Jaya Singapore Hong Kong Tokyo
Nairobi Dar es Salaam Cape Town
Melbourne Auckland

and associated companies in
Beirut Berlin Ibadan Nicosia

Published by Oxford University Press, Inc.,
200 Madison Avenue, New York, New York 10016

Library of Congress Cataloging-in-Publication Data

Emery, James C.
Management information systems.

(Wharton executive library)
Bibliography: p.
Includes index.
1. Management information systems. I. Title.
II. Series.
T58.6.E46 1987 658.4'038 87-7922
ISBN 0-19-504392-8 (alk. paper)

1 3 5 7 9 8 6 4 2

Printed in the United States of America
on acid-free paper

Series Foreword

The Wharton Executive Library is designed to fill a critical need for a new kind of book for managers. The rapidly changing business environment poses a major challenge to senior executives that cannot be met by the traditional strategies that produced yesterday's and today's winners. Technological advances and their diffusion, dramatic changes in the structure of key industries that result from government deregulation, the rising tide of mergers and acquisitions, the change in consumer lifestyles, and the internationalization of business are some of the forces management must contend with.

Modern business schools are aware of these challenges and much current business research is concerned with finding concepts and methods to help managers solve their new range of problems. However, it can take as long as five years for useful academic work to reach managers—progressing through presentations at academic meetings, limited circulation of working papers, publication in scholarly journals, and finally perhaps reaching publication in one of the nontechnical journals directed at practicing managers. Given the nature of the current business environment, this is too long a delay.

The Wharton Executive Library provides executives with state-of-the-art books in key management areas without the usual time-lag. To enhance their usefulness, each book is:

- Up to date; reflects the latest and best research
- Authoritative; authors are experts in their fields
- Brief; can be read reasonably quickly
- Nontechnical; avoids unnecessary jargon and methodology
- Practical; includes many examples and applications of the concepts discussed
- Compact; can be carried easily in briefcases on business travel

Senior managers will find the volumes in the series to be especially valuable. Because of the books' readability, top management can use them to assess whether their key managers are aware of and using the latest concepts in, for example, marketing, forecasting, financial accounting, finance, strategic planning, information systems, management of technology, human resources, and managing multinational corporations. Given the strong interdependency among the various management functions, it is essential that senior executives be aware of the best academic thinking in the areas for which they are ultimately responsible.

The qualities that make these books useful to managers should make them equally valuable as assigned reading in executive development programs in colleges and universities and management training programs in industry. They can also be used as supplementary reading in academic business courses where their state-of-the-art content should nicely complement theoretical material in standard textbooks.

It is appropriate that this new series should originate in the Wharton School, the first business school founded in the United States—and probably in the world. Its faculty is at the top of its profession, and its graduates fill the management ranks at all levels in all parts of the world. The school's academic focus is on the functional areas of management with an analytic and empirical approach. This quality and approach are clearly evident in these books.

We, the series editor and the publisher, hope that you, the reader, will share our enthusiasm for the series and find the advice of some of the leading scholars in their fields, as presented in these books, of value to you.

We also welcome any suggestions and comments from our

readers that will help us to find new and useful titles and achieve the widest possible audience for the books.

YORAN (JERRY) WIND
The Lauder Professor
The Wharton School
University of Pennsylvania
Series Editor

Preface

This is a book about how managers can realize strategic advantages from mangement information systems. It is based on two fundamental premises:

1. The effective use of information technology will become increasingly vital to the long-term success of virtually all organizations.
2. The implementation of an effective information system depends critically on informed leadership by senior managers.

Serving in this leadership role is not easy for many managers. Success demands that a manager participate actively in the implementation process. Sometimes the manager may have to take initiatives that conflict with past practice and conventional wisdom.

Consider the following actual case. About a year ago the chairman and president of a large financial services company began reviewing ways to replace one of the firm's major data processing applications. The program was expensive and difficult to maintain. Worse, its ponderous inflexibility made it impossible to adapt to changing product needs. As a result, the firm had to forego possible business opportunities and expend extra manual effort to make up for the software's deficiencies.

The company examined a number of alternative approaches, from buying an application package to developing entirely new custom software. The price tag of each alternative fell consistently in the vicinity of $10 million. The executives regarded this as an unacceptable price, so they resigned themselves to living with their existing unsatisfactory system.

As a last resort, they considered a proposal from a team of three software developers who thought that they could implement a custom system for about a tenth of the cost of the other alternatives. The magic on which they based their implausible estimate was a high-productivity "fourth-generation" language. Virtually none of the data processing and operational managers of the firm regarded the proposal as credible, and they recommended strongly against accepting it.

What would you do if you were president of the firm faced with this dilemma? Conventional wisdom suggested that the proposed approach offered little chance of success. Unfortunately, though, conventional wisdom offered no attractive alternative. The president chose to accept the innovative but risky proposal.

The story is by no means over, but already signs point to a dramatically successful implementation. After about six months of intensive work—much of which was spent setting up a new computer center and learning how to use an entirely new development language—the software is now at an advanced prototype stage. It includes most of the business functions that the existing system provided, plus a number of new and useful functions that were previously viewed as infeasible. The productivity of the development team is estimated to be greater than ten *times* that of conventional COBOL programmers.

The current prototype version is extremely adaptable. A requested change can usually be handled overnight. Even a major modification or enhancement takes only two or three weeks, in contrast to the months or even years required for similar changes in a conventional system. Senior executives of the firm—including those who were initially most skeptical about the new approach—now anticipate major competitive advantages from being able to implement sophisticated new features quickly and cheaply. It seems very likely at this stage that the president's gutsy decision to proceed along the innovative path will prove to be far less expensive and risky, and result in a far better system, than any of the conventional "safe" alternatives.

This case illustrates the problems and opportunities facing most organizations these days. It also demonstrates the need for top-level visionary leadership. Most organizations are a long way from fully exploiting the strategic opportunities that information technology offers. The gap between practice and potential has probably never been greater, and the penalties for failing to narrow the gap are bound to grow.

With continuing improvements in the power and price of computers, management information systems will play an increasing role at the operational, tactical, and strategic levels of the organization. Through the remaining part of this century, the substitution of low-cost information processing for more expensive resources—land, labor, and capital—will undoubtedly be one of the most attractive opportunities for staying competitive in a global economy.

The stakes involved and the profound changes a truly effective information system is likely to bring about make it impossible for management to stand above the fray. Creating an effective system is inevitably costly and disruptive. Building it requires continued user participation and a willingness to take some risks and make substantial long-term investments. Only senior management can provide the necessary commitment and the shared common vision to achieve success.

This book is designed to give senior managers the knowledge necessary to play their vital role. It does not aim to convert managers into technical experts, but does aspire to provide them with the insights necessary to assess their options. Some managers may be extraordinarily good at judging the advice of technical experts without themselves knowing much about the technology, but buttressing intuition with sound concepts is likely to make a manager a more effective leader in the implementation process.

The argument is often advanced—not infrequently by the technical experts whose advice is being judged—that a general manager need not know much about information technology. The analogy of driving a car is sometimes used to support this view: a driver can use an automobile perfectly well, so the argument goes, without knowing anything about the internal combustion engine.

That is generally the case, of course, because such knowledge is not likely to affect the user's ability to drive from one point to another. Since all automobiles are based on much the same mature technology, and the driver interacts with the machine through a standard and well-known interface, few drivers can expect to gain a competitive advantage through superior knowledge of automotive technology.

A Formula One driver, however, who needs to push engine technology to the limit, could not afford to be wholly ignorant of the technology of internal combustion engines. Likewise, a manager of a large automobile fleet in which performance, efficiency, durability, and maintainability of the engines are significant issues would probably find it quite useful to know the general principles of an engine's operation.

A similar argument applies to management's use of information systems. If the information system is viewed merely as a standardized back-office function that involves little uncertainty or provides few opportunities for competitive gain, then management is quite right to pay it scant attention. If, on the contrary, the information system presents continual nagging problems or offers substantial potential advantages, it is hazardous for a manager to ignore the system and rely solely on technicians to see that it is employed effectively.

It is primarily for the practicing senior manager that this book is written. Managers from a broad spectrum of organizations should be able to benefit from the material, because the basic principles of information systems apply equally to large and modest-sized enterprises, profit-making firms, nonprofit organizations, and government agencies. The book takes advantage of the reader's general understanding of how organizations work, but assumes no background in computer technology.

I hope that technical personnel involved in developing management information systems can also benefit from the material. Finding themselves on a technological treadmill, they may have little opportunity to look at the broader aspects of their work. This book can give them a valuable conceptual framework for communicating with management and developing more effective systems.

I have made every effort to make the book as helpful and "friendly" as possible. My principal aim was clarity in discussing what are often quite complex issues. To this end, I have used examples drawn from a variety of organizations and management functions.

My primary focus is on concepts rather than on specific techniques. In a field that changes as rapidly as information technology, techniques change rapidly but basic principles endure. A typical manager reading this book will be called upon to make difficult computer-related decisions over the next several decades—a period that will witness continued revolutionary changes. Only a sound conceptual understanding of the underlying issues will equip the manager to deal with a changing and unpredictable world. Within the general conceptual framework, I have tried to provide useful guidelines for setting management policies.

I have not shied away from using the jargon of the trade. After all, a specialized term can provide an effective handle for thinking about a complex problem and for communicating concisely with others. The book provides a comprehensive glossary of terms that a reader might find unfamiliar. When each term is introduced in the text it is printed in **boldface.** The book also includes a detailed index and an annotated list of references at the end of each chapter.

I owe a considerable debt to a number of people who have commented on various versions of the manuscript, including Fred Collopy, Dudley Cooke, Frank Palm, Yasuki Sekiguchi, and Giuseppe Traversa. My two colleagues, Uday Apte and Stephen Heffner, with whom I am now writing an expanded version of the book for classroom use, took special pains to review the manuscript. I am also grateful to Christine Fidler, Yan Lau, and Christopher Shull for their support in preparing diagrams and other material. My wife Cornelia served as an ideal model of a nontechnical reader and wielded an unrelenting editorial pencil. Herbert Addison, Executive Editor of Oxford University Press, provided more useful suggestions than an author has a right to expect of a senior editor.

An acknowledgment of my debt to all of these persons does not, of course, constitute an endorsement by them of the ideas expressed—as they would readily attest.

Philadelphia, Pennsylvania J. C. E.
July 1987

Contents

MANAGEMENT INFORMATION SYSTEMS

❖ 1 ❖

The Information Age

The Revolution In Information Technology

The Age of Change

"May you live in interesting times" is said to be an ancient Chinese curse. If change creates interest, then there can be little dispute that we live in an interesting age. Whether we view change as a curse or blessing, we might as well accommodate ourselves to a continual revolution.

Change has always been with us, of course; what is new is its current pace. Since the Renaissance, and especially during the 20th century, we have experienced an accelerated rate of change—in science, technology, medicine, economics, government, philosophy, and art. There are ample signs that the past is a mere prologue to an ever-greater rate of change in the future.

Information technology provides the engine that drives much of the current change. Based on spectacular advances in **microelectronics**, information technology is moving us toward rapid and revolutionary changes. Just as previous technological developments put their stamp on an era, information technology is becoming so central to our economy, our culture, and our daily lives that we are entirely likely to regard the emerging era as the **Information Age**.

Development of Information Technology

Information has been a scarce commodity during most of human history. Written material was extremely limited and expensive, numerical calculations were painfully laborious and practiced by a small fraction of the population, and communications were sluggish and unreliable.

Things began to change with the invention of movable type and relatively inexpensive printing, by Gutenberg and others in the 16th century. Numerical calculations were greatly facilitated by the invention of mechanical calculators, beginning with the work of Pascal and Leibniz in the 17th century. Jacquard in the 18th century and Babbage in the 19th century anticipated punched card storage and automatic programming. Morse's telegraph and Bell's telephone provided the first steps in our ability to substantially reduce geographical limitations in handling information. Hollerith's punched card devices, employed first in analyzing the 1890 U.S. Census, made large-scale data processing practical and widely available. Aiken's development at Harvard of the Mark I calculator in the early 1940s applied the latest electromechanical technology to the task of automatic computing. The foundations for the Information Age had been laid.

Modern Technology

The development of ENIAC, the first all-electronic automatic general-purpose calculator, at the University of Pennsylvania during the mid-1940s, provided a key ingredient in information technology. Free of the mass and time lags associated with electromechanical devices, the electronic computer is capable of an enormous increase in the speed and volume with which information can be manipulated. A **stored program**—automatic control of the computer by means of a program stored within the internal memory of the computer—frees the computer to execute at electronic speeds rather than the relatively glacial pace of a human operator.

Virtually all modern **computers** are based on **digital** technology, in which data are represented using a **binary** (0–1) coding scheme. Electronic devices are especially compatible with binary coding, because a two-value code can easily be represented in the form of two physical states of a device—a punched hole or no hole, a pulse or no pulse, positive or negative current, magnetization in one direction or the other, and so forth.

At the heart of the information processing revolution is the **microelectronic chip** capable of storing and manipulating digital information at staggering speeds and very low cost. For example, the current state-of-the-art memory chip, which is about a quarter-inch square, is capable of storing the equivalent of about 40 single-spaced pages of text at a cost of less than 10¢ a page. The information it contains can be accessed for processing in less than a tenth of a **microsecond** (i.e., one ten-millionth of a second). If progress in microelectronics continues at its historical rate—which seems almost certain—the cost by the end of the century of the microchips needed to store an entire large encyclopedia will be less than $100, and any portion of it will be accessible in less than a billionth of a second.

The computer is one of the major beneficiaries of the revolutionary advances in microelectronics. Like memory chips, microprocessors—"computers on a chip"—have dramatically improved in speed, cost, size, and reliability. These developments have contributed substantially to a continual drop by about 20% per year in the cost of a computer with a given functional capability. This rate compounds into a tenfold cost reduction each decade.

Such a rapid rate of change is not easy to absorb. A tenfold—or *order of magnitude*—advance in an important technology has generally brought about profound changes. For example, the advance from the horse and buggy to the automobile, and from the railroad to the subsonic jet aircraft, both represent about an order of magnitude advance in transportation; each has led to major changes in the way we live and work—some good, some bad, but substantial in any case. In computers we see an equivalent order of magnitude change *each decade.* It is no wonder that we experience problems in deploying such technology.

In the case of **telecommunications**—the transmission of information via electrical signals—we see a rapid conversion from **analog** to digital technology. In an analog telephone network, for example, sound is represented in the equivalent form of a transmitted electrical signal having an amplitude and frequency that correspond to the sound characteristics. With digital technology, sound is represented in numerical form by measuring its characteristics at frequent time intervals. The data can then be transmitted, stored, and manipulated using standard digital technology. Eventually the coded data are converted back to analog form to reproduce the original sound at the receiver's end.

Implications of the Information Revolution

The pervasive rush to digital technology has enormous implications. The computer, the communications network, the copying machine, and the thermostat can all exploit the economy and reliability of microelectronics. More important, they can all be made "smarter"—capable of providing vastly expanded functional capabilities under the control of a stored program computer. One of the most critical of these new functions will surely be connection to a communications network to share resources, exchange information, and provide more integrated control. The common digital representation of information in all the devices greatly facilitates such interconnection and integration.

We are thus rapidly approaching a world of smart, interconnected machines capable of providing a brand new and expanded range of services. This will inevitably have a profound effect on the way an enterprise is organized and managed. Its structure should recognize, after all, a manager's ability to coordinate within and across organizational boundaries. Coordination involves information processing, and will therefore benefit from rapid advances in information technology.

These changes in individual organizations will also be reflected in national and international economies. Information technology has already begun to have a considerable effect in technologically advanced societies. Shifts in work content and the creation of new jobs have resulted in well over half of U.S. employment being composed of **knowledge workers**—clerical workers, teachers, librarians, journalists, financial analysts, computer programmers, and the like. These workers are essentially in the business of dealing with abstract information rather than concrete products like wheat, coal, and automobiles. The industrial worker, like the farmer, may become a small fraction of the work force (which, of course, does not imply that total industrial output will go down; the American farm sector, employing less than three percent of the work force, produces more than ever and is still a critical segment of the economy).

Where will the Information Age lead? It is relatively easy to forecast such important technological variables as the price and raw power of computer hardware: merely extrapolating from past trends will not be too far off. The difficult part is predicting organizational, behavioral, and political effects. We can have little confidence in our ability to foresee the changes that will be wrought by a profoundly different technological world that features a lavish proliferation of powerful interconnected computers.

In the face of relative certainty about the extent and inevitability of technological change, but with great uncertainty as to the detailed consequences of the change, management is left in a quandary. About the best thing it can do under the circumstances is establish a solid organizational competence in information systems and build an enabling technological infrastructure that facilitates adaptation to change.

Information Systems in the Information Age

The Information-Intensive Approach

How is an information-rich world likely to differ from the world of the past? For the most part, our organizations and procedures were established in an era when information processing was expensive and limited in power. Consequently, we learned to manage with sparse information. In the Information Age the economics are reversed: we will seek ways to substitute inexpensive information processing for such expensive resources as labor, cash, inventory, land, plant, and equipment.

The classic justification for computerization has been labor savings in routine clerical operations. Labor saving is still a frequent objective of an information system, but the current approach tends to focus on broader savings in *mainline* activities—the activities that can really contribute to the success of the enterprise because they account for a disproportionate share of the value added by the organization. Some examples:

- In a warehouse operation, all the items included in current customer orders are sorted into bin location sequence so that stock pickers can reduce their travel time in collecting material for shipment.
- The cash management system makes detailed time-phased projections of cash inflows and outflows to minimize unproductive float.
- The truck dispatching system schedules the loading and routing of trucks in a way that minimizes total mileage—saving fuel, labor, and capital investment in trucks.
- Automotive engineers use the computer to perform a detailed simulation of the movement of a vehicle over the road to design effective shock absorbers—saving engineers' time and the cost of producing actual working models to test a design.

These examples have a common theme: the information system helps the organization to work smarter, not harder. The computer provides an abstract analog—a **model**—of the physical world that allows the substitution of information processing for physical resources. As the cost of information processing continues to decline sharply relative to all other resources, further attractive opportunities for such substitution will continue to arise.

In this process, the information system begins to approximate more closely an accurate and comprehensive analogue of the organization and its activities. The end objective is not to eliminate the physical world—after all, a manufacturing company, for example, eventually has to produce real products. A faithful information analogue, however, allows the organization to experiment and plan in the inexpensive abstract world of the computer before execution in the expensive physical world.

Characteristics of a Strategic Information System

An information system that begins to exploit the potential of information processing will no longer be confined to the back-office role of routine data processing. It will become, in fact, a strategic resource that has a major influence on the organization's ability to compete. For information-intensive businesses, such as financial services, competence in managing information has already become critical to the enterprise's success. In other types of organizations—manufacturing, distribution, retailing, government, health care, education—an effective information system will become an increasingly essential ingredient for success or even survival.

Different organizations will follow different paths to success. Each has to choose the **critical success factors** that the organization regards as appropriate for its business, culture, and capabilities. These are the relatively few activities that must be done well to maximize the organization's prospects for success—becoming one of the low-cost producers in its industry, say, or providing the highest quality service to its customers.

A strategic information system concentrates on the functions that contribute to the achievement of the organization's critical success factors. Employees have to be paid and customers billed, so the information system must continue to deal efficiently with conventional data processing functions. A focus on these kinds of applications, however, is not likely to yield a strategic advantage. Strategic advantages are gained when the organization is able to distinguish itself through lower costs, better products or services, or unique capabilities. Some examples:

- A distributor selling to retail stores uses its history of past orders to prepare reports for its customers that analyze demand patterns and generate suggested new orders.
- A company places communication terminals in the offices of its major customers, allowing them to enter orders via the telephone system directly into the supplier's inventory system, thereby giving its customers the benefits of faster ordering procedures and reduced delivery times; this in turn allows its customers to gain better service while running with substantially lower inventories.
- A financial services company analyzes data about past customer transactions to tailor its marketing approach for new services.
- A manufacturing company develops a **CAD/CAM** (computer-assisted design/computer-assisted manufacturing) system that allows it to provide custom-tailored products at competitive prices.
- A bank uses a sophisticated model to provide individualized financial and tax advice to middle-income customers.
- A charitable institution keeps a file of the lifetime gifts of contributors to develop highly individualized relationships with them.
- An organization responsible for managing pension and health care programs allows its clients to analyze and select from a rich variety of alternative benefit plans.
- A transportation firm maintains an instant-access file of shipment status data that provides information for managing operations and supporting customers.
- A consulting company furnishes its principals with powerful interconnected workstations, allowing them to compose and edit their own documents and access a variety of analytical and retrieval services that substantially improve the quality and productivity of their work.

The list could go on, but the point should be clear by now. In each instance, the information system is used to gain an important advantage in an area that matters to the organization. More often than not, the approach has a heavy market orientation, offering unique or enhanced services felt directly by customers in a way that builds long-term relationships. In almost all cases the telecommunications system is used in some way, often through links to parties outside the boundaries of the organization itself.

Coordination of an Organization

Large **organizations**, and many small ones, could scarcely operate as they do now without the aid of computers. Nevertheless, existing information systems devote most of their resources to **transaction processing**—the handling of customer orders, billing, payroll, accounting, and other routine operational matters. Relatively few organizations can legitimately be said to use their information system as a fundamental tool of management. That is likely to change.

Most persons in an organization are engaged in information processing of some sort—generally not formalized or based on computers, but information handlers nevertheless. Managers spend the bulk of their time communicating in meetings, face-to-face discussion, and over the telephone; occasionally they plan and make decisions. Office personnel receive and disseminate messages, type documents, schedule meetings, and search files. Factory supervisors assign work and monitor progress, schedule machines, and look for lost work orders. Like the U.S. Army, most organizations have few workers actually on the firing line: hardly anyone really does anything "productive" like tilling the soil, cutting metal on a lathe, or selling to customers.

Although management may have a legitimate concern that the organization devotes too large a fraction of its resources to activities not directly involved in the output of the enterprise, it is nevertheless true that indirect methods are the hallmark of the advanced technological society. In a primitive economy, a high percentage of the population is directly engaged in agriculture or some other production process; each worker, however, produces relatively little output. In an advanced economy, all the vast machinery, procedures, and overhead of the organization are directed at making the few workers on the firing line as productive as possible.

Management Through the Information System

A comprehensive and effective information system can go a long way toward improving, and even selectively replacing, the communications and coordination part of the organization. Such an information system becomes a critical part of the "central nervous system" of the organization, through which information is collected, transmitted, stored, manipulated, and displayed. Decisions are made on the basis of information stored in the system's **database**—the collection of data available to the system that gives the status of the organization and a history of its activities. Well-defined decisions, such as production scheduling or the disposal of an insurance claim, will increasingly be relegated to decision-making components incorporated within the formal computer-based system.

Needless to say, humans served by the system should retain control. For the foreseeable future they will continue to make the more critical and less structured decisions. Often they will be aided by selective information and models that predict the consequences of alternative courses of action. A good system has ample "dials and handles" by which the user can control the operation of the system—for example, how it filters the huge volume of transaction data to present highly condensed decision-making information.

A good information system permeates all parts of the organization. At the **operational** level, where the organization's work gets done, the computer provides much of the "intelligence" to handle high-volume transaction processing and routine decision making. At the **tactical** level, dealing with medium-term resource allocation, the computer plays a vital role, but the responsibility remains with human decision makers. At the **strategic** level, involving broad and long-term policy matters, the formal information system plays an important but distinctly subordinate role.

Why an Effective Information System Is So Difficult to Implement

There is nothing infeasible or impractical about a comprehensive information system of the type described. Indeed, some leading-edge organizations already have a number of the pieces at least partially in place. Although developing a strategic information system is necessarily a long evolutionary process, beginning the quest is a worthy endeavor.

Most existing systems fall woefully short of what they could and should be. Dealing only with pedestrian data processing, they are not at all aimed at strategic needs of the organization. They generally provide little direct operational support and have only modest impact on decision making at the middle or higher levels of management. They are apt to be inflexible, unable to respond to anything but the most critical and pressing demands—and even then with considerable delay and at a high cost. Despite large and growing expenditures to maintain and improve these systems, their imperfections persist.

Conceptual Difficulty What accounts for these problems? First, developing a truly effective information system is genuinely difficult. It is an inherently complex intellectual task to spell out explicitly in excruciating detail how the organization should be run. It requires participants in the process to deal with unaccustomed abstractions about the organization's values and procedures. Such knowledge must eventually be translated into a form that the computer can interpret. Even under the most favorable conditions, the organization must go through a period of disruption and learning to develop a comprehensive information system.

Technological Gaps Despite the extremely rapid strides made in technology over the past three decades, frustrating gaps still remain. The necessary technology usually exists, but lashing the parts together is not easy. The task is made all the more difficult by a lack of industry standards to facilitate the integration of hardware or software supplied by different vendors. The goal of being able to connect any part to any other part has proven to be illusive.

Rapid Rate of Change Organizations must struggle just to keep up with the rapid rate of change in technology. Many fail. Caught in the expensive quagmire of having to maintain obsolete programs to provide continuity of services, the typical organization has few resources left over to make any fundamental improvements. Its technicians have little time to keep up with advances in the field, and consequently they may not be well equipped to apply state-of-the-art technology to remedy their situation.

Lack of Vision by Technical Personnel Sometimes difficulties arise from limited vision on the part of technical management. Often rising through the technical ranks, managers of information systems activities may lack a sound understanding of the organization's business needs. By overlooking the forest for the trees, these managers may fail to grasp the potential strategic advantages that the organization could gain by focusing on critical business needs. Many common shortcomings of information systems—such as technical elegance at the expense of relevance, hardware efficiency at the expense of flexibility, and obscure human interfaces—stem in part from the difficulties some technical managers have in seeing problems from the perspective of the user and the needs of the business. This can feed on itself, setting up a barrier to communications and cooperation between users and the technical staff.

Lack of Vision by General Management Management's lack of vision is the ultimate source of the difficulty in creating an effective information system. Senior executives are responsible for putting in place the management and resources necessary to install and operate a cost-effective information system. To the extent that the system falls short of needs, senior management must take responsibility.

Many top executives are not comfortable with this responsibility. Unlike most other business functions for which they are accountable, the information system activity seems to them shrouded in unpenetrable jargon and technical complexity. Lacking confidence in their ability to calibrate their information system against a reasonable standard of excellence, they are forced to resort to crude norms, such as a perceived comparison with peer organizations and their industry's average expenditure for information systems as a percentage of revenue.

A major problem is that much of what executives know about computer systems may no longer be true (if it ever was). Having been burned in the past with wildly optimistic statements about how computers would improve their lot, many executives now have fairly low expectations. They have been conditioned to think of computer systems as being overly expensive, difficult to implement, unresponsive to real needs, and largely beneath their personal concern and involvement. It is hard for them to accept the claim that things are different now, and that a computer system can be benign and strategic. They have heard that one before.

Achieving a Successful Information System

It may be difficult to develop an effective information system, but it is certainly not impossible. Achieving success is akin to baking a fine soufflé: it is obviously possible (because many chefs do it routinely), but success is not foreordained (because soufflés often collapse). Virtually any organization should be able to implement a successful information system if it has the will and acquires the necessary skills, but some organizations will continue to fail because they lack these essential ingredients.

By adhering to a few basic ideas, management can substantially increase the likelihood that the organization will achieve success:

- Management should play the primary role in establishing broad policies for applying information technology.
- Although the information system must deal adequately with routine data processing tasks, it should also devote some of its resources to the organization's critical success factors.
- Special attention should be given to new applications that provide direct support of operations and decision making (increasingly through the use of personal computers linked via telecommunications to shared computers and databases).
- The system should be developed in an evolutionary fashion—that is, through a series of relatively small steps that enhance the system's capabilities and adapt to the organization's changing needs.
- The system should be implemented with the use of "fourth-generation" computer languages and other high-productivity software tools that substantially reduce the time and cost of software development and provide the flexibility necessary to permit an adaptive, evolutionary development process.

Management's Involvement in Setting Information System Strategy

Nontechnical senior managers have various attitudes about information technology:

1. It is not important for either the organization or the manager, and therefore it deserves little of the manager's attention.
2. It is important for the organization to do an effective job in applying the technology to routine tasks, but it has little relevance for a senior manager.
3. It is important for the organization and the manager, but its arcane details are beyond the manager's ability to understand.
4. It is, or is capable of becoming, a critical strategic asset well worth the effort for the manager to understand its essential details and get involved in its deployment.

Over the brief history of computer-based information systems, managers' attitudes have certainly moved from indifference to growing concern. It is only quite recently, however, that the general idea has begun to take hold that the computer can become a strategic asset worthy of top management's involvement.

What the Manager Needs To Know About Information Technology

Management's growing interest in information systems still leaves unsettled whether it is feasible for a nontechnical manager to acquire enough background to define information needs and establish general deployment strategies. A well-designed information system must balance its cost against the value of the system. The value comes from how it affects the behavior of the organization, which is primarily a management issue. Cost, on the other hand, is a technical issue, and depends on how the system is designed to meet requirements specified by the users. This means that the persons who establish the general tone and direction of MIS strategy should not only have a very good understanding of how information can enhance the effectiveness of the organization, but they should also have sound insights about the technology and what it takes to apply it successfully.

Technical Material This is not an unreasonable expectation. Providing such a background is, in fact, the purpose of this book. My goal is to present material that a manager should know to play an effective role in the deployment and operation of an information system. I have tried to limit the technical material to the most important matters contributing to this goal.

A few words of explanation are in order concerning a manager's need for sound insights about technical issues. It is certainly true that a general manager need not acquire detailed technical skills to be an effective user of information technology. It is also true that a manager's job with respect to the information system is to deal with policy issues and identify useful ways of deploying the technology. Nevertheless, an oganization's failure to develop a successful information system often stems from management's inability to understand both the power and the limitations of information technology.

Consider, for example, one of the the more critical technical developments, **database management systems**, or **DBMS**. A DBMS is a software product used in managing the organization's data resources. Its use has tremendous implications for the information system's effectiveness, flexibility, and cost. This technology has been evolving for more than two decades and is now widely employed. Most organizations, however, took far too long to grasp its importance, and are still a long way from exploiting its full capabilities. Management's general understanding of technical issues can have a powerful effect, I believe, in positioning the organization to deploy developments of this sort more effectively.

A little knowledge can be a dangerous thing, and there is always the risk that managers might acquire a superficial knowledge of technology without understanding its more subtle implications. Managers might be extremely loath to pit their technological judgments against the experience of their technical staff. But this is not a new problem for the manager; it occurs whenever the organization employs a specialist such as a lawyer, architect, engineer, or market researcher. Effective managers learn how to deal with them by gaining a broad understanding of the field, by asking probing questions, and by calibrating the responses against their own good sense. This is precisely what a manager should aspire to do in dealing with information system specialists.

Other Material It is easy to focus too much on the technology of information systems. To be an effective participant in the development and use of an MIS, a manager must know something about such matters as the principal functional activities within the organization (marketing, manufacturing, etc.), organizational structure, measurement and reporting systems, planning and control systems, organizational behavior, and incentives. These are, of course, areas in which most managers already have a solid background. The design of an effective information system, however, raises a number of new and important issues.

For example, the way in which we design the organizational structure might change significantly because of new capabilities in communications and decision making. In an information system that collects and retains masses of operational data, the measurement and reporting function might look quite different compared to one designed with a primitive technology having limited information processing capabilities. A manager needs to look at these familiar issues from the perspective of current information technology.

Plan of the Book

I have tried to intersperse managerial and technical topics in a way that motivates interest while providing sufficient background for understanding the important consequences of policy decisions. Chapter 2 covers the basic functions of an information system—the collection, transmission, storage, and manipulation of data and the display of the resulting outputs. Chapter 3 then focuses somewhat more narrowly on the computer and the various languages used in describing computational tasks in a way that permits the computer to execute them.

A management information system can be thought of as consisting of two parts (although they may often be closely related): (1) handling routine transactions such as payroll and customer orders, and (2) providing information for decision making. Chapter 4 discusses the role of transaction processing and how it can affect the heart of the organization's business. Chapter 5 looks at the other part of the system, decision support systems. After receiving relatively little attention for many years, decision support systems are increasingly being recognized as a critical component of a strategic information system.

Chapters 6 and 7 deal with the process of developing application software. Chapter 6 discusses the conventional approach to software development, along with the problems often associated with the process. Chapter 7 then examines some of the newer "fourth-generation" approaches using high-productivity development tools. It is critical that managers understand the substantial changes now occurring in the development process, and the implications these changes have on the successful development of applications.

An MIS is designed and deployed in the expectation that it will provide certain economic benefits. Its development and operation have associated costs. A manager needs to have a clear understanding of the characteristics of the MIS that influence both the costs and the benefits of the system, and how he or she can contribute to the proper balance between them. Chapter 8 discusses these issues.

Chapter 9 looks at some important ideas that provide some practical concepts for attacking the problem of developing a comprehensive information system. By viewing the organization and the MIS as systems, some useful insights can be gained for dealing with such issues as centralization versus decentralization of the organization, and integration versus independence of the MIS.

Drawing on the previous material, Chapter 10 discusses the preparation of an enabling plan and technical capability with which the organization can build a strategic MIS. The aim of the plan is to provide a process and a road map for implementing an information system that meets the broad needs of the organization. Before such a plan can be implemented, however, a great deal of work must be done to provide a suitable infrastructure that allows a successful system to be built and adapted to changing needs.

Finally, Chapter 11 examines the factors that affect the competitive standing of an organization. Methodologies are discussed for assessing the role that the MIS can play in enhancing the firm's position. As always, stress is placed on the necessity for management to lead the way in defining the organization's critical needs and for providing the resources to get the job done.

Further Readings

Davis, Gordon B. and Margrethe H. Olson, *Management Information Systems*, McGraw-Hill, 1985. A comprehensive and scholarly discussion of MIS, with an excellent bibliography.

Lucas, Henry C. Jr., *Information Systems Concepts for Management*, McGraw-Hill, 1986. A contemporary introductory textbook.

Synnott, William R. and William H. Gruber, *Information Resource Management*, John Wiley & Sons, 1981. A management-oriented discussion of MIS with a practical flavor.

❖ 2 ❖

Components of a Management Information System

Introduction to Management Information Systems

What Is a Management Information System?

The term **management information system**, or **MIS**, has been around for a long time. Even now, though, there is no general agreement as to its meaning. One can find support for any of the following points of view:

1. The MIS deals with only the management- or decision-oriented parts of an information system.
2. The MIS involves only the transaction processing part of the system.
3. The MIS combines both decision making and transaction processing components.
4. Lack of agreement on its meaning has rendered the term vacuous; other terms, such as "information resource management," should replace it—thereby escaping from the disrepute into which the term has fallen because of excessive claims by early MIS enthusiasts.

The broadest meaning—the combination of decision making and transaction processing—will be used in this book. We need *some* term for the collection of all formal information processing within the organization, and *MIS* is at least as good as other candidates. Each word is meaningful and significant: an MIS consists of interrelated components—a **system**—that process *information* used in the *management* of an organization.

Using a broad definition of an MIS is not motivated by a mere academic concern with terminology; a holistic view of information processing is, in fact, an important concept for anyone trying to understand the use of information in organizations. Decision makers should be able to draw upon data that enter the system through transaction processing. Their resulting decisions should then be implemented through subsequent transaction processing. All these activities take place within the broad boundaries of the MIS.

This does not imply that everything in an MIS should be tightly integrated. On the contrary, the degree of integration should take account of complex tradeoffs that balance the advantages and disadvantages of independence versus integration. But the design of the MIS should not begin with any arbitrary separation between decision making and transaction processing.

The ongoing digital revolution has led most organizations to expand the scope of their information system well beyond its traditional boundaries. It is becoming increasingly common for the executive with corporate information responsibility—the so-called **chief information officer**—to expand into voice and video communications (as well as data communications), office automation, electronic mail, printing and copying, access to external information services (such as econometric databases), and even the library. There are good reasons to view these information-intensive activities as a whole and to manage them accordingly.

In view of the changing role of information processing within organizations, the following definition of an MIS is not unreasonable: *An organization's management information system is the set of functions that should be included within the purview of its chief information officer.* This regrettably circular (and mildly facetious) definition at least has the merit of putting a concise label on the things with which this book is concerned. The points of view presented will certainly be in accord with a holistic view of an organization's information processing activities.

Structure of an MIS

A comprehensive MIS for a large organization is one of the more complex artifacts routinely constructed by humans. It is composed of a great many parts, or **subsystems**, that interact with one another in varying degree. The structuring of these parts has both a vertical and horizontal dimension. Figure 2-1 represents such a structuring for a manufacturing firm.

Vertical Structuring In the vertical dimension, the MIS has hierarchical layers that deal with matters in varying degrees of detail. The **operational** level handles routine procedures connected with such activities as processing customer orders, preparing shipping documents, and invoicing a customer. It is at this level that the bulk of the high-volume data processing takes place. Close links are maintained between the operational system and the physical processes carried on by the organization.

The operational subsystems collect data about events in the real world—the arrival of a customer order, say. Data entering the MIS are stored temporarily in a data depository, called the **database**. They remain there until further processing takes place, such as handling the shipping and billing functions associated with a customer order. In the course of this processing, the database gets adjusted to reflect the consequences of the event (a reduction in available inventory by the amount of the customer order, for example). The output of the processing may be a working document of some sort (e.g., a shipping label or a customer invoice) or, increasingly, it may be displayed in transient form on a TV-like screen.

The **tactical** level of the MIS deals with relatively short-term decision making, such as production scheduling or the grant-

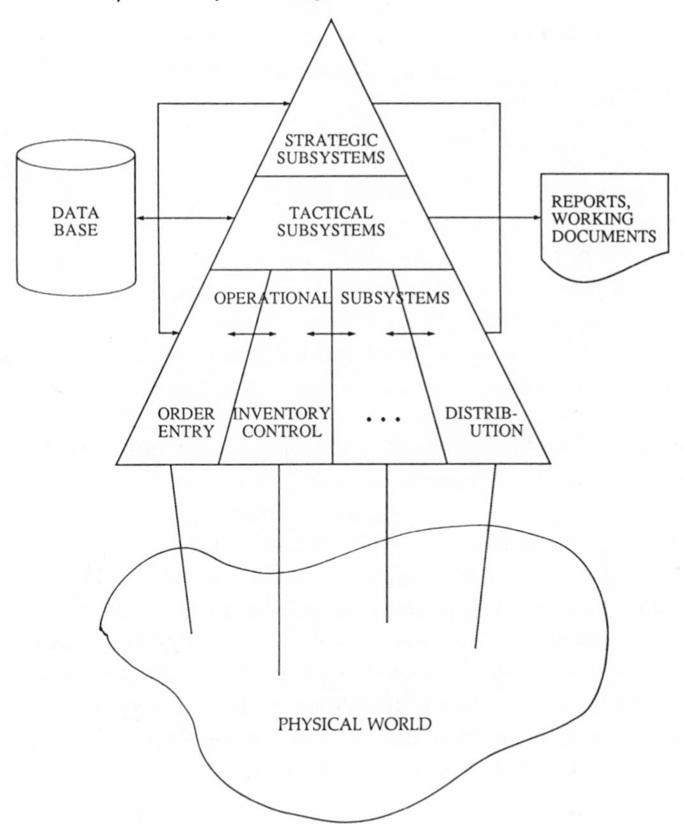

Figure 2-1. A management information system.

ing of a bank loan under general policy guidelines. Unlike operational functions, in which much of the activity is handled automatically by the computer, tactical decision making generally involves the direct participation of a human. Much of the information to support this decision making comes from the database, and almost always requires some form of computation to summarize the detailed operational data or to make projections using a forecasting technique of some type. Other data to support tactical decision making may come from formal external sources (an econometric database giving forecasts of gross national product, say) or from informal sources (e.g., casual reading of the *Wall Street Journal*).

Decisions made at the tactical level generally get implemented through the operational part of the MIS. A production schedule, for example, might be implemented by preparing the detailed working documents that accompany physical material as it gets processed through the various production stages; in a more sophisticated system, the equivalent information could be displayed on a terminal at each production site in the shop. In an **integrated** MIS, the information necessary to link a tactical decision with the operational level may be exchanged automatically through the database; more often, though, the information is passed between the levels by a more informal means (by giving a printed copy of a production schedule to the production supervisor, say).

The **strategic** level is similar to the tactical level, except that it deals with broader, longer-term decisions such as a product development plan or a major corporate acquisition. The boundary between tactical and strategic components is not at all distinct: one blends into the other. At the higher levels in the hierarchy, decisions become longer term, broader, and less susceptible to formalization. As a result, they rely less on formal information in the database and instead depend heavily on informal sources of external information.

Horizontal Structuring Within each level, functions are broken down into **applications**. The operational level in a manufacturing firm, for example, would typically include an **order entry** subsystem (the part dealing with the processing of customer orders) that feeds data to the inventory control, production scheduling, distribution, and accounting applications.

In some cases these subsystems may be directly coupled to one another, providing a high degree of integration. More often, the applications exchange information indirectly through a shared database or they rely on independent sources of data. As we will see in later chapters, the degree of integration among subsystems is a major design issue.

Basic Information-Processing Functions in an MIS

Despite the overall complexity of the MIS, the functions performed within each subsystem tend to be conceptually straightforward. Data enter the system, and are then transmitted, stored, manipulated, and displayed. Because these functions serve as fundamental building blocks for the MIS, it is useful to examine each of them in some detail.

Data Entry

Data entry serves as the sensory function for the MIS. To keep in touch with the physical world, the system has to cope with a constant barrage of data about real-world events. These data enter in the form of **transactions**, which describe events such as the receipt of a customer order, the sale of an item at a supermarket counter, the completion of a manufacturing operation in the shop, or the arrival of a check from a customer.

Data Entry Techniques A variety of means are used to enter data. The traditional method is through a manual keyboard operation, which is still the dominant approach in most systems. Virtually all new systems use a terminal device in which the keyboard is attached to a TV-like screen (often called a **CRT** for cathode ray tube, or—**VDU** for video display unit). After the operator has identified to the computer the type of input transaction to be entered, the computer displays a blank **form** that shows the operator which pieces of data, or **fields**, are to be entered. An example of such a screen in shown in Figure 2-2, in which the rectangles show the data to be entered by the operator. Each such field is associated with a field name (e.g, John Jones is associated with the NAME field), permitting visual verification. On command of the operator, the entered data on the screen are stored for later processing.

Important advances have been made in **optical scanning**

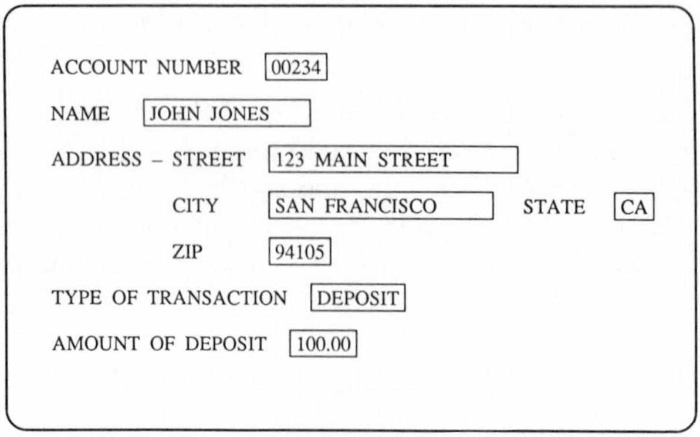

ACCOUNT NUMBER [00234]

NAME [JOHN JONES]

ADDRESS – STREET [123 MAIN STREET]

CITY [SAN FRANCISCO] STATE [CA]

ZIP [94105]

TYPE OF TRANSACTION [DEPOSIT]

AMOUNT OF DEPOSIT [100.00]

Figure 2-2. Data entry screen.

technology, which provides automatic "reading" of printed characters, product identification codes (e.g., the Universal Product Code, or UPC, used in supermarkets), or even handwritten characters. Direct automatic sensing of real-world conditions, such as flow rates or temperature in a processing plant, can be appropriate in some circumstances. This is particularly true if the application involves automatic **real-time** decision making, as in a process control system for an oil refinery in which a decision must be made within a limited reaction time of the ongoing physical process.

Cost of Data Entry The data entry function is one of the most critical parts of the MIS. For one thing, it is expensive—often one of the larger components of cost, in some cases accounting for the bulk of continuing operating expenses. In fact, the cost of data collection in a given situation can often be so great that the value of the information is not worth its cost. In the retail industry, for example, collecting transaction data on individual items sold may be so expensive that less precise means have to be used, such as estimating retail sales on the basis of the bulk delivery of merchandise to the store from a warehouse.

The development of sophisticated scanning devices has substantially lowered the cost of data collection. With the introduction of this technology, it has even become economically feasible to collect item sales data at the checkout counter of a supermarket, which is characterized by having high volume, low value per unit, and thin profit margins.

Error Control The data entry function is critical not only because of its cost, but also because it is one of the most pervasive sources of errors in the system. The common adage "garbage in, garbage out" (GIGO) recognizes that the quality of the outputs of an MIS depend on the quality of its input data. Obtaining accurate data is a difficult and never-ending task.

Accuracy is generally achieved through some form of redundancy. The traditional means is through a **verification** process, which involves a comparison of the data entered with the original source of the data (a handwritten sales order, for example). This is sometimes done through a complete re-keying of the data, followed by an automatic detection of any discrepancies between the first and second keying. With the widespread use of CRTs, visual verification has become more common.

An exceedingly important aspect of error reduction is an **editing** program in which various automatic checks are applied to the data to determine their accuracy and completeness. In editing the data to establish an employee record for a newly hired person, for example, the computer might perform the following error checks:

- The name of the employee is checked to see that it consists only of alphabetic characters (to guard against, say, the erroneous entry of the digit zero in place of the letter O).
- The Social Security number is checked to see that it has the format NNN-NN-NNNN, where each N is a digit.
- The date of birth is checked to determine that it falls between a "reasonable" range of dates (after 1920 and before 1980, for example).
- The year of the employee's graduation from college is checked to verify that it exceeds the date of birth by at least 12 years.

As is perhaps evident from these examples, the computer can subject data entering the system to all manner of thorough and complex error checks. The greater the likelihood of a given type of error, or the worse the consequences if an error occurs, the closer the data entering the system should be scrutinized automatically by the system. As the probability or cost of an error declines, however, there eventually comes a point at which the cost of an error is less than the cost of avoiding it.

Integrated Data Entry In developing an application, designers have a variety of approaches that can be taken to reduce both the cost and error rate of data entry. The most fundamental improvement comes by reducing the volume of data to be captured. This can be achieved by collecting a given piece of data only once and thereafter sharing it among all applications that require it. Sales data might be collected by the order entry system, for example, and then be made available to the inventory control, production scheduling, and accounting systems. This sort of data sharing is one of the most important characteristics of an integrated MIS. Integration does not come without cost—it adds to the complexity of the system and the amount of coordination needed to develop and run it—but the tradeoffs these days almost always favor substantial data sharing, at least for high-volume inputs.

Interactive Data Entry Collecting data as an integral part of normal operations offers great opportunities for improving data entry. At the operational level we are seeing a growing use of **interactive** terminals—that is, terminals linked directly to a computer in a way that allows a dialogue between user and computer. In a manufacturing shop, for example, a supervisor interacts with the system to obtain production schedules and inform the system of work that has been completed. In a claims processing office for an insurance company, the claims agent might deal with customers face-to-face or over the telephone, entering the appropriate data during the course of the discussion. The agent can obtain information on the status and coverage of the customer's policy, initiate the claims handling process, and advise the customer about any further steps required.

Many organizations are actively pursuing a plan to convert the bulk of their applications from the traditional **batch processing** mode to the interactive mode. With batch processing, transactions are collected over a period of time (a day, say) and then processed in a single batch—usually long after the job was completed in the shop or the policyholder left the claims office. With interactive processing, the computer is available to provide information and support for operational personnel. These aids can substantially improve the efficiency and quality of operations.

Similar benefits are gained from the standpoint of data entry. Since an interactive terminal is connected to the computer during the course of entering a transaction, the computer is able to guide the process (e.g., show the claims agent what information is required for a given type of claim) and perform immediate error checks. If an error is identified by the computer, the terminal operator can generally correct the error immediately with very little extra effort or delay (unlike batch processing, where error correction is a major problem and source of delay). Because operational personnel maintain close touch with real world activities, they are in a much better position to correct errors than data entry specialists who know little about operational matters.

A subtle but exceedingly important advantage of direct data entry by operational personnel is the motivation it provides for high accuracy. If an error is made in entering a transaction, operational staff personnel must generally deal with the consequences. They therefore have a strong incentive to enter data correctly in the first place, and correct them rapidly if an error is detected. It is almost impossible to provide comparable incentives in a system with less direct coupling between operations and data entry, and consequently the problems of preventing and correcting errors tend to be considerably greater.

Thus, in addition to all its other operational advantages, an integrated interactive information system contributes substantially to efficient and accurate data collection. As a result, this design is rapidly becoming the favored approach for most new operational systems.

Data Storage

A dominant characteristic of a management information system is its heavy reliance on stored data. The system must maintain vast files of data to provide the information for transaction processing and decision making. In a hospital, for example, a five-day period of care for one patient might easily generate 25,000 characters of data pertaining to such matters as the patient's medical history, laboratory reports, medical treatments, and billing. In a hospital with an average of 300 occupied beds, this would give rise to well over 500 million characters of stored data per year.

Role of the Organization's Database The collection of stored data constitutes the database of the MIS. The database provides an analogue of the real world, which is then used as the basis for assessing current conditions, analyzing past events, generating plans for future actions, and controlling execution of the plans. To the extent that the organization is managed through the formal information system, the analogue world described in the database is a substitute for reality. Organizations are certainly moving in the direction of greater reliance on the MIS for operational, tactical, and even strategic planning and control, and so it is crucial that the database be kept a sufficiently faithful representation of reality.

A variety of different types of data are stored in the database:

Status data, which describe conditions as of a given point in time (e.g., number of units in inventory, pay rate for each employee, patient status).

Transaction data, describing past events (receipt of a sales order, hiring an employee, admission of a patient).

Summary operational data (e.g., sales over the past year, broken down by month, product line, and geographical location; hospital bed occupancy rates by week over the past five years; average operating times and scrap rates in a manufacturing process).

Engineering data (engineering drawings, bills of material for each manufactured product, chemical processes).

Textual data (correspondence, business reports, published documents, bibliographic references, interpersonal messages).

Images (pictures, graphs, signatures).

The database is kept current through continual **updating** as part of transaction processing. When a transaction enters the system for processing, the computer must retrieve related data from the database. At the conclusion of the processing, the computer stores updated data that reflect changes due to the transaction. When a customer order is received, for example, the database must be updated to show the amount of the sale, an appropriate reduction in available inventory, an increase in the customer's accounts receivable balance, and possibly changes in production schedules and purchase requisitions. Virtually every transaction leaves a "footprint" of some sort in the database.

Organization of the Database The database is not just a pool of unstructured data; it has to be organized in a way that facilitates access to selected parts.

The basic component of a database is a named collection of data called a **file**. The personnel file, for example, might have the name EMPLOYEE, and contain such information as the name, address, date of birth, education, and employment history of each employee. A file can be referenced by its name and operated on in various ways. The command COPY EMPLOYEE NEWEMPLOYEE, for example, might be used to create a backup copy of the EMPLOYEE file, with the new copy being assigned the name NEWEMPLOYEE. Other commands can be used to move the file to a different physical location in storage or delete the file altogether.

In most cases a file consists of a homogeneous collection of **records**. The EMPLOYEE file contains employee records, for example, and the CUSTOMER file contains records on customers. Each record generally pertains to a single entity of interest to the system—a given employee or customer, say. Each record must have a unique **identifier** through which it can be accessed in the database—the employee's Social Security number in the case of an employee record, say. Records within a file can be added, deleted, rearranged (e.g., sorted in SSN sequence), and modified in various ways in the process of updating the file.

In some cases a file may not have an internal structure for accessing its individual components. A document, for instance, could be stored simply as a long string of text. Any processing in this case would generally operate on the entire file—for example, copying it or transmitting it to a specified group of recipients. If portions of the document are commonly wanted, it would usually pay to provide an internal structure through which a desired component can be accessed (by breaking it into named chapters, sections, and paragraphs, say).

Physical Representation of Stored Data The database is typically stored on a combination of physical storage devices. By far the most common medium for bulk data storage is a magnetic surface on which data are coded in binary form. Binary codes are represented by physical states that can take one of two possible values—a spot on the surface magnetized in one of two possible directions, for example.

It should be mentioned, in case the point is not obvious, that users and even most technicians can remain completely unaware of, and indifferent to, the particular binary coding used by their computer. The coding is "wired in" to the hardware, so an automatic correspondence exists between a given character and its binary coding. If a user types the letter A on a keyboard, for example, the binary code for that letter is automatically generated and retained within the system. Any subsequent display of the character will show the original letter A. The subject of binary coding is brought up here principally for the sake of dispelling any mystery about data representation within an MIS, and because the terms bits and bytes have to some extent joined the popular jargon.

For purposes of discussing binary coding, we can use a "0" to represent one physical state and a "1" to represent the other (although the magnetized spot has no physical identification with a zero or one, of course). Each two-state binary number (i.e., 0 or 1) codes one **bit** of information. A collection of eight bits is generally called a **byte**.

The scheme used in binary coding of textual material simply establishes a consistent and unique (but essentially arbitrary) correspondence between a given combination of bits and the character being coded. With a one-bit code, two different characters could be coded—for example, "A" corresponding to a "0", and "B" to a "1"; with a two-bit code, four characters could be represented (e.g., A, B, C, D) corresponding to the two-bit combinations of 00, 01, 10, and 11. In general, with n bits we can code 2^n characters.

Consider a coding scheme using one byte, or eight bits, to represent each character. This provides a total of 2^8, or 256, different combinations of eight bits (00000000, 00000001, . . . , 11111110, 11111111). With a character set of this size we could therefore code all of the upper and lower case alphabetic characters (including a variety of alphabets associated with different languages), the ten digits, and a large number of punctuation marks and special symbols (e.g., . , " ? / ! @ # $ £ ¥ % &). Such a character set is quite sufficient to represent the typical business record consisting of text, numbers, and special symbols.

Graphical or pictorial information can also be stored in binary form, but not through the use of a limited set of coded characters as is done with conventional text data. A common method is to form the picture from a series of dots (or **pixels**, for picture elements). A moderately high-resolution CRT screen, for example, can display about 1000 dots along each of its two dimensions, requiring a total of 1000 × 1000, or one million, pixels to form an entire picture. The picture is represented by defining each of the million pixels as being turned either "on" or "off."

In the case of a monochrome display, a pixel can be coded with only one bit, since each has only two possible states; it thus takes a million bits to code a single screen's worth of pictorial information having a 1000 × 1000 resolution. These bits are stored in the internal memory of the computer, taking 125K bytes of storage (since each of the eight bits of a byte can be used to code one pixel). This information is sometimes displayed through a **bit-mapping** process in which each bit in memory is mapped to a specific location on the face of a CRT screen or on a printed page.

For a multicolor representation of the same resolution, in which any pixel can assume one of, say, 16 different possible colors, each dot requires four bits ($2^4 = 16$), or a total of four million bits per picture. This is equivalent to about 75,000 words of character-coded information, suggesting that the old adage that "a picture is worth a thousand words" substantially understates the case. If finer color differentiation and higher resolution are required—256 colors with a 2000 × 2000 resolution for publication-quality pictures, for example—storage requirements grow by a factor of eight, to 32 million bits. Although digital-coded pictorial information accounts for a very small proportion of most existing databases, its use is bound to grow as the MIS expands into non-traditional areas.

Physical Storage Devices Two types of storage media currently dominate all others in terms of the volume of data maintained. The traditional high-volume medium, and still very common, is **magnetic tape**. The binary codes for the data are stored sequentially along the length of the tape, analogous (except for the coding used) to the storage of sounds or video material on a magnetic tape. To process data, the data must be **read** from the tape into the computer, where the actual manipulation of the data takes place. After updating the information, the computer **writes** it onto another tape to prepare an updated version.

Reading the full length of a magnetic tape takes a minute or more. This **access time** from tape occurs at a leisurely pace compared to the internal processing times of the computer, where a transaction might be processed in a small fraction of a second once the computer gets the necessary raw data. Further compounding the delay, the reels or cassettes of magnetic tape on which the data are stored are typically stored **off line** in a tape library not accessible to the computer. Before the data can be accessed, a computer operator must manually mount the tape on a tape drive. Once mounted, the computer then

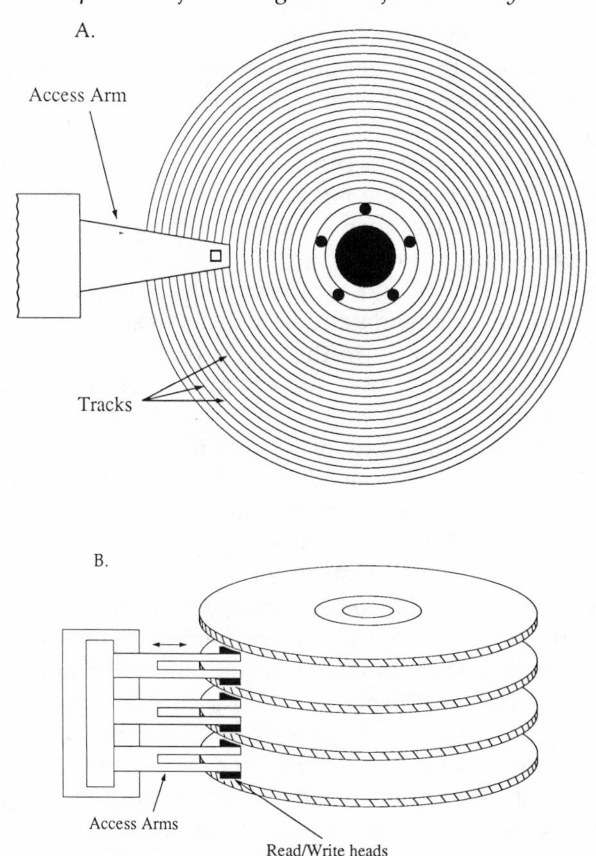

Figure 2-3. Construction of a magnetic disk. (A) Recording surface. (B) Access mechanism.

has control over reading from, and writing to, the tape, with no further human intervention required.

The most rapidly growing segment of the database is the portion retained on **magnetic disk**. Here the data are recorded on the surface of a continuously spinning round disk, somewhat analogous to the way music is recorded on a record except the data are stored in concentric rings, or **tracks**, instead of a spiral groove, as shown in Figure 2-3A.

A disk drive may have one or several disks, with recording typically on both the upper and lower surfaces of each disk. A given record is accessed by moving a **read/write head** over the track on which the record is stored, as shown in Figure 2-3B. The head is then able to sense the contents of the magnetic recording. The access time for the computer to obtain data from a disk ranges from less than 1/100 of a second (10 **milliseconds**) to 1/5 of a second or more.

Both magnetic tape and magnetic disk are **erasable** media—that is, they allow new data to be written over old data (which are, of course, destroyed in the process). An inventory record stored on a magnetic disk, for example, can be updated by merely writing the new version in place of the old. (To allow for additions and deletions of records in updating a magnetic tape file, a new file is created on a different tape rather than selectively modifying the contents of the old tape.)

A magnetic disk is the most common example of a **direct access** device, so called because the computer is capable of accessing more or less directly any data stored on a surface of the disk, without having to access intervening records (as in the case of magnetic tape, in which the entire tape must be read to get at the last record). Another direct access device, the **optical disk**, uses a recording technology not unlike that used for digital recording of sound on a compact disk. This technology has now reached a practical stage for some application. It provides the advantages of storing an exceedingly high volume of data at a very low cost, while still giving the relatively fast access time associated with a direct access device. Current optical disk technology offers only non-erasable storage, which can be a disadvantage in some cases but a distinct advantage in others (for permanent archives, maintained for legal purposes, for example).

On-Line vs. Off-Line Storage Because the data on a direct access device are accessible relatively quickly and without human intervention, the data are said to be **on line**. An on-line system, then, is one in which the database is stored in this accessible form, allowing the computer to interact with the user (or a physical process controlled automatically by the system) with very short response times. Most systems being developed today rely heavily on on-line storage for retaining the active part of the database in a readily accessible form.

If cost were not an issue, all databases would be retained in on-line storage. By historical standards, on-line disk storage has reached a very low cost—less than $30 for a million characters (a **megabyte** in the jargon of the trade). Even so, this cost is about 100 times more expensive than off-line tape storage. For very large and relatively inactive databases, magnetic tape offers a practical and cost-effective alternative to disk. (Optical storage, if it meets its promise, may offer magnetic disk performance at magnetic tape costs.)

Transaction data show a common pattern of use that allows designers to take advantage of the characteristics of a hierarchy of storage devices. When a transaction first occurs, the data are referenced frequently during the course of editing and updating. Generally a number of subsidiary transactions follow the initial transaction, which give rise to multiple accesses to the transaction data. A sales order, for example, may be followed by a series of transactions to acquire, ship, and bill for the items ordered by the customer. It is important to keep transaction data in an accessible form during this active period of a transaction's processing. On-line storage is increasingly being used for this purpose.

On completion of the processing associated with a transaction, the probability drops sharply that detailed transaction data will be needed. Accordingly, after a few months individual transactions can be expunged from on-line magnetic disk storage and transferred to off-line magnetic tape (or, increasingly, optical disk storage). This provides a cost-effective means for long-term storage of **archival** data—that is, data with a low probability of access, but that nevertheless must be retained in a form that permits subsequent retrieval and analysis (perhaps for legal reasons, as in certain regulated industries).

In place of the detailed transaction data, summary data are usually prepared and retained in on-line storage in a much more compressed form. Sales data, for example, might be summarized into monthly sales by product, geographic location, and industry. This offers the dual advantages of reducing both the storage requirements and the processing time needed to generate useful information (at the risk, however, of washing out some relevant details).

Imperfections in the Database Several sources of distortion cause the database to be a less than perfect analogue of the real world, including the following:

- Errors in data entry (e.g., by recording a shipment from stock of 15 units when it was actually 25, resulting in overstating the inventory position by 10 units).
- Delays in updating the database (e.g., processing shipments at the end of the day, so that during the day the database does not reflect the current day's activities).
- Inconsistencies in related duplicate data due to differences in update timing or data definitions (e.g., in shipment figures coming from sales and manufacturing).
- Failure to collect or retain detailed data, resulting in a database that provides only a "low resolution" description of reality (e.g., the database may not record the current status of work in process on the manufacturing floor).
- Failure to make relevant associations among data segments (e.g., to recognize that checking account customer Frederick C. Hoffman with an overdrawn account also has a $10 million trust account under the name F. C. Hoffman).

All these distortions can be reduced: the system can perform more error checks, update the database more quickly, use common data to avoid inconsistencies, collect and retain more data, and provide rich associations among related data segments. However, perfection is impossible to attain and expensive to approach. One must trade off the value of a more faithful representation of the real world against the cost of achieving it.

Fortunately, advances in information technology make it feasible to maintain the database so that it provides an increasingly detailed and accurate analogue of the real world. Interactive operational systems improve the accuracy and timeliness of data. The continuing dramatic improvements in direct access storage technology permit the retention and ready access to vast quantities of transaction data. Advances in database management systems make it feasible to link related data segments and retrieve comprehensive information on a selected topic (e.g., the status of all accounts that a specified customer maintains with a bank).

Computation

Some form of computation lies behind all the automatic "intelligence" exhibited by the MIS. Computation includes not just the obvious example of arithmetic calculations; it also encompasses data manipulation such as sorting data (e.g., into alphabetical order) or comparing two values and taking different actions based on the result (e.g., handling an inventory backorder if the order quantity exceeds the on-hand balance).

Structure of a Computer System Computation takes place within the **central processing unit**, or **CPU**, of the computer. In performing the computation, the computer is directed by a **program** consisting of a series of steps designed to achieve some desired result (e.g., generating paychecks). The program is stored in the computer's **memory**, along with any data involved in a computation. (The memory is generally called **main** or **primary storage** to distinguish it from **external storage** such as magnetic disk or tape.)

The memory of the computer is one of its most critical resources. Even on very large machines, the memory has a capacity of a few dozen megabytes of data and programs, whereas the database may consist of many thousands of megabytes. The solution to the problem is to keep the entire database on mass storage, bringing into the computer's memory only the very small fraction of the database currently involved in some sort of processing. If, for example, a sales transaction enters the system for processing, the database records for the customer and each of the inventory items on the order would generally be read into the computer's memory from disk storage. Here the actual computation takes place. The results are then written back onto the disk to update the database.

Computer Processing Power A typical large commercial machine executes a program at a rate of 10 to 100 million instructions per second, or **MIPS**; even a relatively inexpensive personal computer (costing $1000, say) can compute at a rate in excess of 100 thousand instructions per second (.1 MIPS). Advances in the speed of computation have continued at a prodigious rate over the past three decades, and show every sign of continuing at a similar (or even accelerated) rate through the remainder of the 20th century. It is entirely likely that at the end of the century almost every manager and professional worker will have his or her own personal computer selling for a few thousand dollars with the computing power of today's mainframe machines.

How can each individual personally soak up the same computational power now generated by the central machine for a fairly large organization? The answer is that there appears to be an insatiable demand for cheap computer power. A number of user support tasks are extremely computationally intensive. Generating a high-resolution, multicolored, moving graphics display, for example, can quite easily keep a mainframe computer fully occupied. With such a capability, a plant manager could display the results of a detailed simulation of a manufacturing facility as work flows through the shop. It will take at least an equivalent computational capacity to provide reasonably general voice recognition capabilities as a means of facilitating a user's interaction with a computer.

Information Display

The computer must have means to communicate with the outside world to gain access to input data and to make available the results of its computation. It does this through various **input/output devices**. A personal computer, for example, typically uses a typewriter-like keyboard as an input device and a CRT screen and printer as output devices. In the case of a larger computer shared among multiple users, inputs generally come through a communications network that links a variety of remote terminals. Outputs, too, may be distributed through the network, to be stored or displayed at a remote site, or they may be stored or printed locally.

The display function of an MIS provides an essential link, or **interface**, between the system and the user. The purpose is to present information in a way that enhances the user's ability to perceive and act on the facts conveyed by the information. The effectiveness of this interface is becoming all the more important as we move rapidly toward the (almost) universal use of fast-response interactive systems that depend heavily on a close rapport between user and machine.

Options Available A designer has a variety of options in displaying information; among them are the following:

- Printed (or **hard copy**) form versus a transient display on a CRT screen.
- Presentation in graphical form versus text or tabular form using character data (letters, digits, etc.).
- Monochrome versus multicolor.
- The level of resolution of graphical output (i.e., the density of the pixels that form the picture).
- The output rate of the display device (e.g., number of characters per second printed or the response time to generate a bit-mapped screen).
- The features for manipulating a graphical CRT display (e.g., the ability to rotate a 3-D diagram or to "zoom" to a larger or smaller image).
- The level of quality of printed characters (e.g., fully-formed **letter quality** printed character versus characters formed with a series of connected dots on a **dot matrix printer.**

Almost any combination of these characteristics can be chosen for a given application. Substantial progress has been made over the past years in improving the features, quality, and cost of display devices. The trend has certainly been toward higher resolution, multicolor, multifunction devices for both hard copy and screen output. For example, the forthcoming generation of CRTs for widespread use with personal computers is likely to provide a rich selection of colors displayed with a resolution in the neighborhood of 1000 points in both the horizontal and vertical dimensions. For hard copy output, a very attractive technology is the laser printer, which provides near-typeset quality (300 dots per inch or better in each dimension) and allows a combination of character and graphical presentations.

Despite these important advances in display technology, the real problem remains the design of the human interface so that the system provides the most effective way of presenting information to users. We run some considerable danger of using fancy technology to display trivial, irrelevant, or incomprehensible information. A common manifestation of this problem is the use of gloriously colored "exploded" pie charts for displaying simple one-dimensional information that might better be printed in the form of a simple table.

Part of the difficulty in presenting information is that perception is a very idiosyncratic thing. Some users are good consumers of numerical information presented in graphical form, while others prefer a tabular format. In using a graphical display, there are a limitless number of alternative presentations. Choosing the most effective one—or, more realistically, even an acceptable one—requires more art than science.

It is virtually impossible to know in advance which presentation best meets the needs of a given user. The design almost always requires an adaptive trial-and-error process in an actual working environment. As a way of conserving development and processing costs, the conventional methodology has been to place the primary emphasis on finding a single standard display format that adequately meets the needs of a wide range of users. Fortunately, the new technology now makes it feasible to develop tailored and adaptive outputs for each individual user.

Communications

Why Communicate? Newly emerging information systems differ most strongly from those of the past in their growing reliance on communications. To serve as the organization's central nervous system, the MIS must tie things together through communications. A need for communications arises throughout the enterprise:

- Information enters the MIS through numerous interactive terminals scattered around the organization (and even outside, on the premises of customers, suppliers, government agencies, and the like); such information must then be communicated to other points in the system where transaction processing takes place or where action is taken on the basis of shared information.
- Remote computation services or data are needed to process a transaction or inquiry at a local site.
- An expensive or specialized resource, such as a high-resolution laser printer or graphical plotter, is shared with multiple users by linking their personal workstations to the shared resource—ideally in an accessible enough form to give each user the illusion of having his or her own dedicated resource.
- Person-to-person communication is needed for planning and coordination.

Communications can take many forms: letters, face-to-face meetings, or the physical delivery of a storage medium (a reel of magnetic tape or a floppy disk, say). The primary concern these days, however, is with **telecommunications**—the transmission of information by electrical means. The transmission medium may be wire, microwave signal, or (increasingly) fiber optics. For reasons already discussed, the strong trend is toward digital technology in which all information (including voice and image) is transmitted in the form of discrete binary codes.

Historical Developments The current situation facing MIS designers can best be appreciated by a brief review of earlier communications technology. The history of advances in management information systems is closely related to advances made in telecommunications. Early systems—those developed in the 1950s and much of the 1960s—depended very little on electrical transmission of data, because telecommunications technology was complicated, expensive, slow, and unreliable. Because the need for communications existed then as now, data were principally communicated by physically transporting the storage medium on which the data were recorded. Paper documents (sales orders, say), punched cards, or reels of magnetic tape were delivered to the computer center, where they were then further processed. The printed outputs from the processing—invoices, paychecks, management reports, and the like—were delivered to the designated recipients. This approach suffered from the time delays associated with the physical delivery of inputs and outputs, but it was quite compatible with the technical and economic tradeoffs of the day, as well as with the inherent lags of the batch processing mode that dominated information systems at that time.

The next step in the evolution of communications came in the mid 1960s with the widespread use of **remote job entry (RJE)** terminals. An RJE terminal typically had an input mechanism capable of reading a deck of punched cards, with a printer as its principal output device. The terminal was linked through a communications line to a shared central computer, where the actual processing took place in batch mode. This arrangement permitted users located near the RJE terminal to transmit their input data and receive their printed outputs without the inconvenience and delay of physically transporting them. Although a welcomed improvement, RJE terminals certainly did not offer a completely satisfactory solution. For one thing, the terminals were expensive enough (generally $25,000 and up, in 1960 dollars) that they generally had to be shared by dozens of users scattered over a considerable area (an entire building, say). Furthermore, the system was intrinsically batch oriented, and therefore did not provide responsive operational support for users.

Relatively low-cost communications terminals also became available during this period. They allowed an individual user or small workgroup to communicate with a shared computer. The typical terminal for this purpose used a keyboard as the input device and either a CRT or a low-speed typewriter-like printing mechanism for output. These terminals were quite "dumb"—they incorporated no computer-processing capabilities, and they could store data locally in only the slow and inconvenient form of punched paper tape; all computer intelligence and most storage capacity were provided by the shared central machine. A user could enter input data and receive a more-or-less immediate response (depending primarily on the current load on the computer). Alternatively, the inputs could be stored temporarily in a batch queue at the central site and then processed at a convenient time.

Distributed Systems As hardware prices continue to decline, it has become attractive to add more "intelligence" and storage capacity at the user's **personal computer,** or **workstation**. Virtually all workers—managers, professional, clerical, and operational—will eventually have their own workstation available to provide interactive support, stand-alone personal computing, storage of selected portions of their own private database, and access to all network facilities.

A contemporary workstation has, by historical standards, a great deal of computing power. It is capable of executing instructions at a rate of perhaps one MIPS and has a primary memory of a couple of megabytes. Most have one or more attached disk drives that can store in the range of 50 or more megabytes. Furthermore, many have attached disk drives that use a removable flexible (or **floppy**) magnetic disk as the storage medium, giving whatever capacity the user is willing to handle in a personal off-line disk library.

Although most personal workstations are currently used in **stand-alone** mode (i.e., not drawing upon the resources of other computers), it is becoming increasingly common to tie them into a telecommunications network for interpersonal communications and resource sharing. In most organizations these workstations will probably be connected to medium-sized minicomputers maintained at the workgroup or departmental level. This facilitates the sharing of resources among users who tend to interact closely with one another and therefore have a lot to gain from close coordination. It is not uncommon to find, for example, that most of the communications within an organization occurs be among departmental and workgroup associates, while the remaining part goes outside the local organizational boundary (e.g., to higher-level or lateral units).

The advantages of resource sharing can be extended beyond departmental boundaries. Most organizations find it worthwhile to maintain some common resources at the corporate level (or at an intermediate divisional level in a large organization) to avoid uneconomic duplication of resources within multiple lower-level units. This generally calls for the central machine to provide a wide range of generalized computing services.

A so-called **mainframe** computer is usually employed in this central role. Although no precise definition exists for mainframe computers, they tend to have a common set of characteristics. A contemporary mainframe computer must have considerable processing power (10 MIPS and up), large on-line disk storage capacity (often thousands of megabytes), powerful generalized software, and a large technical staff to support its operation.

In a **distributed** computing system, which is likely to become the dominant architecture for MIS, computers throughout the organization are linked through a common telecommunications network. Each remote computer on the network generally has significant stand-alone computing capabilities to serve the specialized needs of its local users, but it also provides access to resources maintained at other locations. The aim is to take advantage of the simplicity and economy of specialized local computers, while also providing users access to more powerful general resources located elsewhere on the network.

A distributed system of this sort provides the vehicle by which the MIS reaches out to all parts of the organization. At the operational level, individual workstations support data entry and transaction processing, providing error checking and intelligent assistance for the staff. For decision making at the tactical and strategic levels, they provide stand-alone computing and access to selective information stored in the user's personal database.

Required resources not available on a personal workstation can be accessed through the network. Most of these are likely to be provided at the departmental or workgroup level where sharing is fast, convenient, and efficient. Resources that are too expensive to duplicate at the departmental level can be maintained on the corporate mainframe computer.

Through telecommunication it becomes quite feasible to reach out beyond the boundaries of the organization, to the premises of customers, suppliers, government agencies, and others. This approach offers tremendous potential benefits in increased efficiency and effectiveness by permitting closer coordination among external parties with which the organization has strong mutually dependent relations. The resulting close ties with customers and suppliers can provide an important means for an organization to gain a strategic competitive advantage.

Network Design Rapid technological changes and regulatory uncertainty make the design of a telecommunication network particularly difficult and risky. Designers face a host of technological issues and alternatives, among the more important of which are the following:

Communication capacity: Each link on the network can handle a certain maximum rate of transmission (called the **bandwidth** of the link, which can be expressed in bits per second, or bps); this can vary from less than 300 bps to over 100 million bps, with costs increasing (but less than proportionally) with increased bandwidth and length of the link.

Network topology: The topology of a network is defined by the way the points (or **nodes**) on the network—that is, workstations, computers, printers, and other resources—are connected; Figure 2-4 shows, for example, four different topologies to connect six different nodes.

Communications standard: For one device to communicate with another, there must be agreement on such matters as the coding scheme used to represent information, the identification of the sender and receiver, and the process of setting up and terminating a communications session (analogous to a telephone conver-

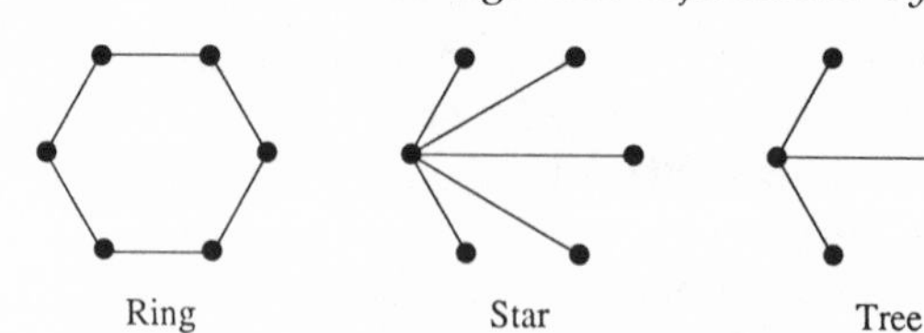

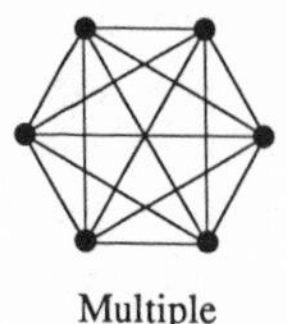

Figure 2-4. Alternative network topologies.

sation beginning with "hello" and ending with "good-bye"); the standard selected by the designers to define these issues is called the network **protocol**.

Leased line vs. public network: For communication links outside the user's local plant or building area, a choice must be made between obtaining the full-time use of private lines (leased from one of the communications companies, such as AT&T in the United States or the public telephone & telegraph provider in other countries), or using a public network.

Telephone network vs. message-switching: If a public network is used, a choice must be made between the regular telephone network in which a communication link is established on an ad hoc (or **dial-up**) basis each time a message is transmitted (as we typically do when we place a personal telephone call), or a message-handling network—also called a **packet-switching** or value-added network (**VAN**)—that transmits individual electronic messages (e.g., credit card transactions from widely dispersed retail stores) in a manner analogous to letters handled by the postal network.

Designers of a network for a given organization must choose a specific combination of network topology, bandwidths, communications protocols, terminal equipment, and communications suppliers. Their objective is a design that meets the particular needs of the organization at an acceptable cost. In doing this, they must consider such factors as the number and geographical distribution of workstations and computers connected to the network, the expected volume of traffic among the nodes, the permitted queuing delays over shared communication links, and reliability and security requirements. Unfortunately, most of these factors tend to change continuously as the system evolves, adding to the design difficulties.

The design and operation of a large telecommunications network are enormously complex and challenging tasks. A lot is at stake. The cost of operating the network consumes an increasing fraction of a growing MIS budget, especially when voice traffic is included. More important still, the organization's well being—perhaps even its survival—increasingly depends on efficient, fast, reliable, and secure communications to tie together its operations and decision making. Few if any issues concerning the MIS have greater strategic importance.

Evolutionary Development of a Communication Network It is impossible to predict with any precision how distributed networks will evolve over the remaining part of the 20th century. It is quite likely, though, that they will have something like the following architecture:

- Colleagues within a department or work group will be connected to a local shared processor through a **local area network** (or **LAN**) that serves a limited geographical area and provides relatively high bandwidth (at least a million bits per second) at a relatively low cost per workstation (as low as a few hundred dollars, but often nearer $1000).
- Each local area network will be tied through a **gateway** connection to a public or private network capable of providing long-distance communications, with the gateway device (a specialized computer) handling any conversion of speed, coding, or protocol necessary to effect inter-LAN communications.
- A corporate mainframe computer center will serve as a central **data hub** for sharing common data, a means of accessing specialized software, and as a source of large-scale computing services.
- Computerized control mechanisms within the network will include the "intelligence" to keep track of the characteristics and location of each network resource to relieve users of this burden.
- The network will integrate communications in all its forms—text, graphs, images, video, and voice—using all-digital coding schemes.
- The computer intelligence built into the network will allow a wide range of powerful new communication functions, such as

electronic mail, voice mail, call forwarding, access to public databases, and (conceivably) natural language translation (between English and French, say).
- Great emphasis will be placed on flexibility in modifying the configuration and level of activity of the network to keep it responsive to an organization's changing needs.

Figure 2-5 shows the basic structure of a network that might evolve over the next several years in leading organizations. It ties together dispersed work groups in offices, factories, and laboratories, and provides a link to the corporate computer center. Each local area network is connected by means of a gateway to the "backbone" network.

Few organizations can take on the enormous task of developing such a network on their own. Most must wait until many of these functions become available commercially as part of the standard offering of communication companies. There is currently a great deal of interest on the part of these companies in developing an **integrated services digital network**, or **ISDN**, aimed at achieving many of the characteristics needed to support a distributed MIS.

Important Characteristics of the MIS

After discussing the principal information processing functions, it is appropriate to close this introductory chapter by amplifying some earlier comments about the more important characteristics of an MIS. These characteristics have significant implications that affect the way in which the system should be designed and operated.

Hierarchical Structure

The MIS, like the organization it serves, is an example of a system. And like the organization, the MIS is composed of a

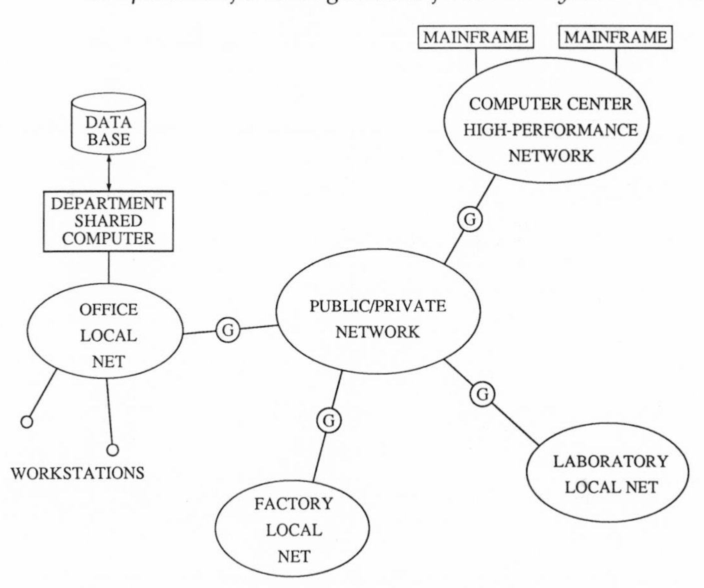

Figure 2-5. A representative telecommunication network.

hierarchy of interacting parts. The choice of structure—how the various parts or subsystems fit together—is an extremely important design issue.

What gives rise to the **hierarchical structure** of the MIS? The system in its entirety is far too complex to implement in a monolithic way. Instead, it has to be broken down into major subsystems. Since each such subsystem is itself too complex for direct implementation, it must be further broken down into simpler pieces. This process continues until the lowest-level component, or **module**, is simple enough to implement without further breakdown.

A key concept in managing the implementation process is to isolate each subsystem enough to gain simplicity, while at the same time deal with the interactions that exist across subsystem boundaries. One of the objectives in choosing a structure is to include closely related tasks within the boundaries of the same subsystem, while separating tasks that have little or no interactions.

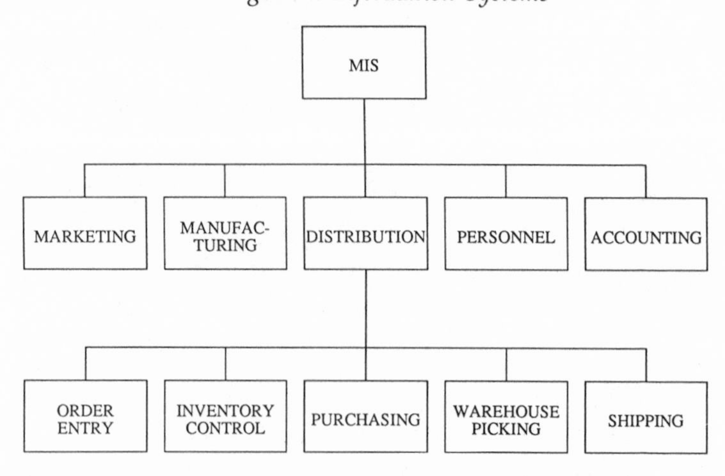

Figure 2-6. Hierarchical structure of an MIS.

Take a specific example. Figure 2-6 shows a representative MIS divided into subsystems for marketing, manufacturing, distribution, personnel, and accounting. These subsystems, in turn, are broken down into lower-level subsystems—the distribution subsystem, for example, into order entry, inventory control, purchasing, warehouse picking, and shipping subsystems. This structure makes sense if each of the subsystems is relatively self-contained. The distribution subsystem, for example, contains components that interact fairly strongly among themselves—order entry certainly has relatively strong interactions with inventory control, for example—but have weaker interactions with the other subsystems at the top level (marketing, for example). Detailed tasks that have especially strong interactions among themselves are combined within a given low-level subsystem (the order entry subsystem, say).

Obviously, the design of an MIS must accommodate important interactions that cross subsystem boundaries. For example, we would certainly want to communicate between the distribution and marketing subsystems. It is critical, however, to maintain close control over any such link, or *interface*, and to keep the number of interfaces to a practical minimum. Too many direct links among subsystems adds greatly to the system's complexity and problems of coordination. (As we shall see in later chapters, *indirect* links through a shared database greatly facilitate coordination among subsystems.)

An issue that comes up in choosing a structure is the degree to which the structure of the MIS should match the structure of the organization itself. On the one hand, much can be said in favor of a fairly close correspondence. For one thing, the organizational structure, like the MIS structure, should be designed to keep closely related activities within the same organizational subunit. The information processing within these subunits tends also to be fairly self contained, and therefore organizational boundaries are often good information subsystem boundaries as well. This close relationship also simplifies client relationships and responsibilities. If the manufacturing vice president, for example, is the unambiguous client for the development of a manufacturing information system, then policy direction tends to be much simpler than if responsibilities are shared with other clients (the VPs of marketing and engineering, say).

On the other hand, the considerations that lead to the choice of a given organizational structure do not necessarily carry the same weight in designing an MIS structure. Geographical proximity, for instance, is often an important organizational criterion, but it can be largely ignored in a telecommunication-oriented MIS. The best structure for the MIS might very well consolidate tasks that previously were scattered among several organizational subunits. The order entry subsystem, for example, might include customer contact functions taken from Marketing, inventory control functions from Distribution, production scheduling from Manufacturing, and billing from Accounting. Although such a structure adds to the problems of dealing with multiple clients, the tradeoffs might favor the more integrated approach.

Distributed Information Processing

As already discussed, most organizations are witnessing a very rapid dispersion of computers. If functional needs call for it, we can afford to scatter computer intelligence profusely, tying the processors together through a telecommunications network. This leaves unsettled, however, the question of how we will choose to exploit new opportunities offered by a distributed environment.

The degree of distribution of processing functions will vary considerably among different organizations. The optimum design depends on complex tradeoffs that consider the nature of the specific applications to be implemented, the geographical spread of activities, and the organization's philosophy of management. These factors will lead some firms to distribute the bulk of information processing away from the central facility, while others might choose to retain most activity at the central site and assign responsibility to the distributed processors primarily for data entry and display.

Distributed computing offers a number of advantages, including the following:

- If a personal computer has the computing power to accomplish a relatively simple task, it generally can do it at a lower cost than a larger machine.
- If a local processor can support interactive applications without continuous access to a remote central mainframe, the costs of communications can usually be reduced.
- A local processor dedicated to a specialized task can generally give faster and more consistent response time than a remote central machine, due to the avoidance of communications delays and queuing time at the shared mainframe.
- Problem solving is generally easier when a relatively simple task is assigned to a separate computer with limited specialized capabilities.
- Programming can be greatly facilitated by high-quality software development tools avilable at modest cost on small computers.
- Expansion of a distributed application can often be done in relatively small, low-cost increments of additional processors.
- The close proximity of a local processor dedicated to local users often has a favorable behavioral effect on users' receptivity and support for the system.

Shared central machines will certainly not go away; they will remain an important part of almost all systems. Such machines offer important advantages for certain applications:

1. Closely integrated applications that share many resources (primarily data).
2. Complex programs that require massive resources—computing power, main memory, and high-volume access to a huge database.
3. Situations calling for close centralized control over shared data.
4. Programs requiring software not available (at an acceptable cost) on mini- and microcomputers.
5. Programs requiring purchased software that is priced on the basis of the number of machines on which it is operated (thus giving an incentive to share the software from a single machine).

One of the great advantages of a distributed architecture is that it does not force an either/or tradeoff: each individual task can be analyzed to determine if it should be distributed or not. Consequently, a good system will almost always be a hybrid of both centralized and distributed computing. Large, complex applications for which the tradeoffs tend to favor centralization will be retained on the central machine; medium-complexity applications will tend to move out to distributed minicomputers at the departmental and work group level; and relatively simple, idiosyncratic applications will gravitate to microcomputers (generally personal workstations). Figure 2.7 shows the

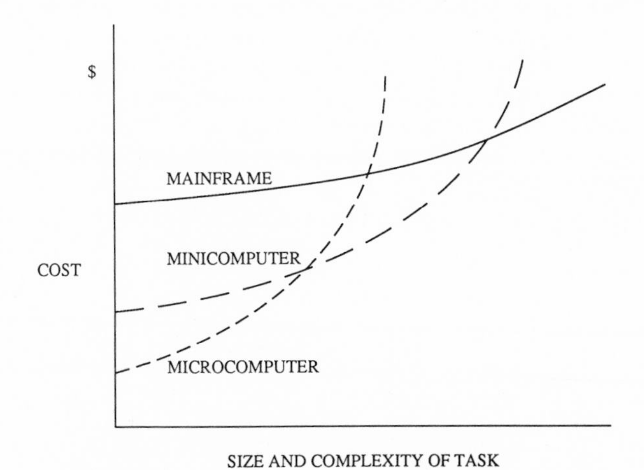

Figure 2-7. Cost as a function of task complexity.

general cost characteristics that lead to such an allocation of tasks among micros, minis, and mainframes.

The development of an effective distributed system of this sort will challenge even the best organizations. Users must play the dominant role in defining requirements and taking responsibility for their own applications. The technical MIS staff has the responsibility for developing a congenial environment for users to get their work done. Such an environment requires effective management of shared databases and communications facilities, high-quality technical support for users, and a responsive set of standards that allows users to develop applications compatible with the rest of the system.

Further Readings

Senn, James A., *Information Systems in Management,* Wadsworth Publishing Company, 1987. A very thorough introductory textbook, with good discussion of design and technical issues.

Tricker, R. I. and Richard J. Boland, *Management Information & Control Systems,* John Wiley & Sons, 1982. A sound discussion of MIS concepts.

❖ 3 ❖

Computer Hardware and Programming Languages

MIS implementers face the task of translating information requirements into a form that can ultimately be executed on a computer. Tremendous strides have been made in developing various **programming languages** and tools that make this task considerably less formidable than it has been in the past. The new languages are beginning to dramatically reduce the cost and time needed to develop new applications. These improvements, in turn, alter in fundamental ways how we approach software development.

The essential approach in developing a **high-level programming language** is to build "intelligence" into a computer **translator** to reduce the burden on the human user. This allows the human to focus on specifying the desired end result, relying on the automatic translator to convert the specification into an equivalent detailed procedure that the computer can execute.

As in all other matters connected with the implementation of an MIS, the choice of a computer language involves tradeoffs. Exercising a high degree of intelligence within a language translator consumes significant computing resources. The translation process itself takes computer time, and the program that results from the process may not be as efficient in the use of machine resources (primarily processing cycles and memory) as a hand-crafted program written in a much more detailed language. As computing technology advances, however, tradeoffs increasingly favor the substitution of computer intelligence for other resources—in this case, the programmer's time.

Because of the close relation between computer hardware and programming languages, this chapter will amplify some of the comments about hardware made in the previous chapter, and then focus on the critical issues of programming languages.

Computer Hardware

Technological Advances

Based on the astonishing rate of progress in microelectronics, computer hardware continues to improve sharply in performance and price. Computer memory, for example, plummets in price by about 30% per year. At this rate, memory that cost $1 million at the beginning of the commercial use of computers in 1954 will cost less than $3 at the end of the 1980s. An entire computer system, composed of more than electronics, drops in price by "only" 20% per year, yielding a cost of less than $400 at the end of the 1980s for a machine having the same power as a $1-million machine in 1954.

Even now, we can generally treat the cost of computer intelligence as having a secondary influence on design tradeoffs compared to all the other costs involved in implementing an MIS. We are rapidly approaching an era of "zero-cost logic" in which the added cost of incorporating computer intelligence within a device is almost negligible. This allows us to include such power anywhere we choose if it adds value in the process. We can see this effect already in numerous home appliances—for example, in inexpensive intelligent phones, cameras, and TVs.

It is important for senior managers to understand the implications of zero-cost logic. It certainly does not mean that MIS budgets will decline. On the contrary, advances in information technology will continue to present attractive opportunities for substituting information processing for other resources; as a consequence, the demand for information services is very likely to increase more rapidly than the decline in the price of computing power. Not only will the total MIS budget increase, but even the hardware component will probably grow in absolute amount (while declining in relative share).

The most important effect of low cost computer power is not on MIS budgets but on MIS design. No longer forced to conserve on computing, designers can embed computer intelligence throughout the information system. The lush use of computer power will be manifested in many ways: the increased intelligence with which the computer handles transactions, compute-intensive human interfaces, powerful support of human decision making, and sophisicated languages that ease the task of expressing our information needs.

Mainframes, Minis, and Micros

As we have already seen, computer hardware comes in a wide variety of machines that differ in cost and capabilities. Although no clear-cut boundaries exist between the different classes of machines, it is convenient to label the high end of the spectrum **mainframe** computers, the middle range **minicomputers**, and the low end **microcomputers**. (The so-called **supercomputers** lie at the highest end of the spectrum in terms of raw computer power, but they are currently used almost exclusively for scientific and technical applications.)

A mainframe is priced in the range of about $250,000 to $5 million and more. It offers a very wide range of features to handle large, high-volume jobs. At the top of the line, these computers can execute millions of instructions per second—25 MIPS and more, and growing all the time. Its main memory may have a capacity of over 100 megabytes, and it may manage a database consisting of more than 50 billion bytes (i.e., 50 **gigabytes**).

A large mainframe computer is almost always used as a shared, general-purpose machine. It usually serves as the central machine at the corporate level or within a major division of a large enterprise. Its workload often includes corporate applications and jobs from operating units that need the power of a mainframe. It generally involves a mix of both batch and interactive computing.

The worldwide mainframe market is dominated by machines produced by the International Business Machines Corporation. Within the United States, IBM's chief mainframe competitors (Unisys, Digital Equipment, and Amdahl) are estimated to share less than 20% of the mainframe market. The market tends to be more evenly divided in Europe and Asia, but even there competitors usually share no more than half of the market with IBM.

Minicomputers are middle-sized machines that sell in the range of about $50,000 to $250,000. They generally have a raw computing power of 1 to 10 MIPS, with main memories of 1 to 16 megabytes. They are often used as general-purpose shared computers for relatively small organizations, such as small companies or departments in larger organizations. Because of their relatively modest cost, they can be dedicated to a fairly narrow specialized application such as the support of computer-assisted design (CAD) or office automation. Unlike the mainframes designed for efficient batch processing, the hardware architecture of minicomputers is targeted primarily toward on-line interactive applications.

The minicomputer market is fiercely competitive, with considerably less concentration than for mainframes. The principal U.S. players are IBM, Digital Equipment Corporation, Hewlett-Packard, Wang, and Data General, with a host of other smaller companies dividing the rest of the market. In Europe and Japan, one or two dominant national companies tend to split their domestic market with the major U.S. vendors. DEC, with a common architecture that spans a broad range of capacities, has been capturing a growing share of the worldwide market.

Microcomputers reside at the low end of the computer spectrum. A practical beginning machine costs around $2,000, but they can range up to $20,000 or more for computationally intensive applications such as high-resolution graphic design. They execute instructions at a rate of about .1 to 3 MIPS, and have main memories of about 250,000 bytes (250 **kilobytes**, or 250KB) to 4 megabytes and more.

A microcomputer is most frequently employed as an individual personal computer. In this role it may be applied to a variety of computing tasks, such as **word processing**, **spreadsheet models**, graphics, and database management. In a small firm a microcomputer may provide the sole source of computing, running such applications as general ledger, accounts receivable, accounts payable, order entry, and payroll.

Although most microcomputers currently are used in a stand-alone mode, the trend is to tie them to a telecommunications network. In this role a microcomputer serves as a multifunction personal workstation offering stand-alone capabilities while also providing a window to the broader range of shared services on the network.

The continuing advances made in microelectronics and computers are reflected throughout the full range of mainframes, minis, micros. Each is growing in power at the upper range, and dropping in price at the entry level. There is an inexorable blurring of the distinctions between the classes of computers, with micros looking every day more like powerful minis, and minis offering more and more of the capabilities of mainframes. This is particularly evident at the boundary between upper-end workstations and minicomputers, where leading vendors offer machines at micro prices that have identical architectures and similar capabilities as their more expensive minis.

The Components of a Computer

All computers, from the smallest microprocessors to the largest mainframes, have the same basic structure. They differ widely in their speed, capacities, and range of special features, but they all have the same fundamental components and operate in the same way.

The heart of the computer is its **central processing unit (CPU)**. It is here that the actual computation takes place, such as adding two numbers or comparing one customer's name with an-

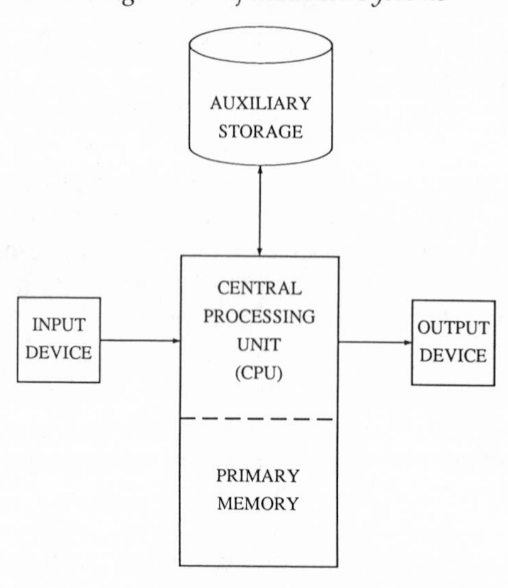

Figure 3-1. Components of a computer.

other in the process of sorting the customer file. All the data used in these computations are retained in the computer's **main memory** (also called the **primary memory** or **primary storage**).

In addition to the CPU and main memory, the computer has to have a means of obtaining inputs and generating outputs from its processing. Input data can be entered from special input devices, such as a keyboard, or can come from a telecommunication connection. Outputs can be displayed on a printer or CRT, or they can be transmitted over a network. The bulk of the inputs and outputs, however, tend to come from, or be stored in, a mass storage device, such as magnetic disk or tape. Figure 3-1 shows a simplified version of a computer.

Program Execution

The CPU performs its computations under the direction of a **program**. A program is designed to accomplish a well-defined piece of work, such as processing customer orders or generating the payroll. It consists of a number of **instructions**, each of which accomplishes only a small task, such as a single arithmetic operation or the movement of a piece of data from one location in memory to another.

Practically all programs contain many **loops**, each of which consists of a series of instructions performed repetitively. For example, a financial analysis program might contain a loop to calculate the **net present value** of a cash flow occurring in a future time period, which would be repeated once for each of, say, 36 monthly periods. Because of this repetitive characteristic, a program consisting of only a few thousand instructions might require the execution of many millions of calculations.

The common measure of computer performance, MIPS, refers to the number of millions of the detailed instructions that the computer can execute each second. It is a fairly rough and simplistic measure, however, because the complexity of work accomplished by an instruction varies considerably among machines. Each mainframe instruction, for example, tends to perform a more complex or productive task than those on a micro, and so the mainframe needs to execute fewer instructions to perform a given processing function.

All the data processed by the CPU are obtained from the computer's main memory, and results from a computation are stored back in memory. The memory is divided into relatively small fragments, called **words**. A common word size for a modest-performance microcomputer, for example, is 16 bits (i.e., two bytes), while most minicomputers (and a growing number of microcomputers) use 32 bits (four bytes). Numbers or strings of characters that exceed the word size (such as the name BARTHOLOMEW CHRISTOPHER FITZPATRICK III, which would require ten words of four bytes each) simply use multiple adjacent memory locations.

Each memory location has a numeric **address**, which gives its physical coordinates. As far as the programmer is concerned, though, the address merely provides a unique identifier for the contents of a storage location. For example, an instruction causing the numeric contents of memory location 10540 to be added to register A in the CPU might be written as ADD

A,10540. (This is a considerably simplified description of what really goes on, but preserves the essential truth.)

The concept of a **stored program** lies behind the generality and flexibility of the computer. Rather than having the logic of a program permanently wired into the computer hardware, a program is stored in contiguous memory locations in the same form as data (i.e., using a binary representation). The behavior of the computer can be completely altered merely by replacing one program with another. If the executing program handles customer orders, we have an order entry machine at our disposal; if the program executes a spreadsheet model, we have a spreadsheet machine.

Like any other data, a program can be read into main memory from external storage (almost always from magnetic disk), from which it can then be executed. The computer can thus move quickly from one task to another, dealing with a very wide range of application programs.

In a **multiprogramming** system, the computer executes several programs more or less simultaneously. Each program is typically kept resident in main memory, so switching from one program to another can occur at electronic speed. The computer typically executes a program for a short period of time, until it needs to access external storage. While this relatively slow operation proceeds, the computer can then switch to another task. A 20 MIPS machine with a disk access time of 10 milli-seconds can, for example, execute 200,000 instructions during the time it takes to make one disk access.

The CPU has a special **instruction address register** that keeps track of the address in memory from which the next instruction is to be executed. The instructions are stored in sequential memory locations (generally one instruction per word), so the computer normally merely increments the instruction address by one for each instruction executed. The exception to this general rule occurs in the case of a **branch** instruction, which allows the computer to break its normal sequential execution of instructions, depending on the particular data values it encounters.

For example, in an order entry system, the computer must be capable of differentiating between the normal condition in which stock is available, and a **stockout** situation in which the order quantity exceeds the available stock. If a stockout is detected for a given item, the computer branches to the program segment that handles that particular condition. This ability of a computer to branch on specified conditions gives the computer much of its power.

Computer Software

System Software vs. Application Software

Computer **software** consists of the various programs that control its operation. The term "software" is derived, of course, from the fact that a program is merely information, an abstraction rather than something concrete and "hard" like the computer itself. There are two basic classes of software: **system software** and **application software**. They differ not in their intrinsic nature (since they are all computer programs), but in the generality of the functions they provide.

System Software System software performs common functions that are not specific to a given application such as payroll or inventory control. The most important examples of such software are the following (each of which is discussed in more detail later):

- An **operating system** manages the general operation of the computer—scheduling work, allocating computer resources (such as main memory), bringing programs into main memory when they are scheduled for execution, communicating with the computer operator, managing files kept in mass storage, managing telecommunication traffic into the system, etc.
- A **database management system (DBMS)** deals with such matters as the retrieval of selected information from external stor-

age and the protection of the database from destruction or unauthorized access.

- A **language translator** translates a program written in human-oriented programming language (such as **COBOL** or **Fortran**) into the machine language that the computer executes directly.

Designers of system software seek to include computational functions common to a variety of specific applications. Making the functions available from the system software permits an application programmer to draw upon these services without having to program the tasks on an individual application-by-application basis.

An excellent example of this approach is the set of generalized database management functions provided by a DBMS. Virtually all application programs have to deal with access to a database, which is an extremely complicated process. Without having generalized database management functions available through the DBMS, application programmers would have to spend a great deal of their time programming these functions for each application.

System software is often produced by the hardware manufacturer, particularly in the case of a computer's operating system. Sometimes this system software comes **bundled** with the hardware at no additional cost, but the trend is clearly toward **unbundled** software for which the user pays extra. In addition to the hardware vendors, independent (**third-party**) vendors produce a variety of widely used system software.

Application Software An application program deals with a particular user function, such as payroll or engineering design. Unlike systems software that provides a generalized set of services for multiple organizations, application software is typically tailored to the specific information processing needs of a given organization.

The traditional way of developing application software is through the use of an in-house staff of **analysts** and **programmers** who prepare programs satisfying the organization's specific needs. It is now quite common, however, to purchase prewritten **program packages** from software vendors. These are usually installed on the client's general-purpose internal computer, but the following alternative means are also available for delivering application processing services:

- An application program may be purchased as a **turn-key** system—that is, a system purchased from a single vendor as an integrated package of both hardware and software.
- Application processing services may be obtained from a **service bureau**—an independent organization that provides computational services to its clients, generally on a batch processing basis (either using a message service for the collection and delivery of input and output documents or a communications terminal on the customer's premises).
- **Time-sharing** computing may be used, in which interactive terminals on the customer's premises are linked through telecommunications to a shared external computer operated by the time-sharing vendor.

Although application packages are designed to serve the common needs of a variety of users, the best ones offer sufficient generality to permit a client to tailor a program to meet many specialized needs. A generalized payroll package, for example, includes a number of options for dealing with such matters as different local and state tax rates, employee deductions, and labor distribution reports. The particular options and rates for a given client can be specified through the choice of a set of *parameter* values, which govern how the program will handle that client's processing. Despite such built-in generality, however, a package often requires some additional tailoring that calls for changes in the program.

Operating Systems

A great variety of tasks must be performed to get useful work out of a computer. For example, even in the relatively simple case of a spreadsheet program used on a personal computer with floppy disk storage, the following functions must be performed:

Prepare a floppy disk to be able to store a spreadsheet model.

Retrieve a file to execute a model.

Print a hard copy of a spreadsheet.

Copy or delete files on the floppy disk.

Compare the contents of one floppy disk with another.

Determine the contents of the files stored on a floppy disk.

Place a "time stamp" on a file, recording the date and time that the file was created.

All these services are provided by the operating system, even though the user may not always be aware of that fact because many of the services are well integrated into the spreadsheet language.

In a large mainframe computer, the operating system provides an enormously complex and varied set of services. At any instant, hundreds of programs and transactions may be residing in the system in various states of completion. The operating system serves as the traffic cop for the entire system, performing the following functions:

- Managing the arrival of new work from input devices and the telecommunications network.
- Storing work in transient queues, waiting for the availability of processing resources.
- Scheduling jobs on the basis of their required resources and priority.
- Assigning hardware resources to jobs, such as allocating a chunk of main memory to a job ready for execution.
- Loading a program into main memory from external storage in preparation for its execution.
- Storing outputs in queues for printing or communication over a network.
- Creating, deleting, or duplicating files.
- Providing security protection for the system, so that no one can easily gain access to unauthorized programs and data through error or criminal intent.

- Handling any error condition that arises (e.g., a hardware failure, program bug, or unavailability of a needed resource).

The transition from a single user running a single program at a time (as on the typical microcomputer), to a multiprogramming system with multiple users and programs (as on a mainframe or shared minicomputer) naturally calls for a tremendous increase in the size and complexity of the operating system. Much interest currently focuses on **parallel processing** as a means of increasing the capacity and efficiency of a machine. An operating system capable of managing a **multiprocessing** environment of this sort is necessarily very complex.

All these powerful services come at a price: a great deal of **overhead** is typically required. The chief overhead resources are main memory and processing time. Major portions of the operating system must reside continuously in memory, ready to take over control of the computer when operating system functions are invoked. In providing such services, the computer must execute the appropriate portion of the operating system program, which absorbs some of the processing capacity of the CPU. It is not uncommon for the operating system of a mainframe to take half of the machine's memory and computing capacity. Some fraction of these resources would be required even if the computer were dedicated to a single program, but the bulk of the overhead resources are devoted to managing shared facilities, with all the complexity that sharing entails.

Language Translators

The computer is useless without a program. The program that sits in memory to control the computer's operation is termed a **machine language** program because it consists of the individual binary-coded instructions that the machine executes directly. Such a program suffers from two serious limitations:

1. Although the circuits of the computer have no difficulty in interpreting machine instructions (since that's what they were wired to do), a human finds them all but incomprehensible.

2. Because each instruction performs very little work, a program that accomplishes a complex processing task typically contains many thousands of instructions.

As a consequence of these limitations, it is extremely difficult for a human to write a program in machine language. Creating and testing each individual instruction is difficult, expensive, time consuming, and prone to error; when multiplied by the large number of instructions in a typical program, the task becomes overwhelming.

Fortunately, it is extremely rare for a programmer to write a machine language program, because there are almost always much better alternatives. This applies even to highly specialized programmers, and it is certainly true for the average application programmer or user.

The insight that allowed humans to escape from the drudgery of machine language programming was the recognition that it is not necessary for the programmer to define the logic of a computing task in the language that happens to be wired into the computer's circuits. Instead, we need only to translate a program written in a human-oriented language into an equivalent program expressed in the machine's own language. Upon execution, the machine language version controls the process in such a way that the computer performs the task specified in the human-oriented language.

Of course, this still requires a programmer to define the task unambiguously in a language that can eventually be translated into machine language. The programmer usually does this by creating a detailed, step-by-step **procedure** (the term **algorithm** is also used) that transforms inputs into desired outputs (processing a customer order to produce shipping documents and an invoice, say). This is not an unknown concept to most of us; an all-too familiar example is a tax form that defines the steps for transforming inputs (the taxpayer's annual earnings, taxes already paid or withheld, deductions, etc.) into an output (the amount owed to the government). A cookbook recipe similarly defines a procedure for converting specified ingredients into a desired end product.

Execution of a program invariably involves the two-step pro-

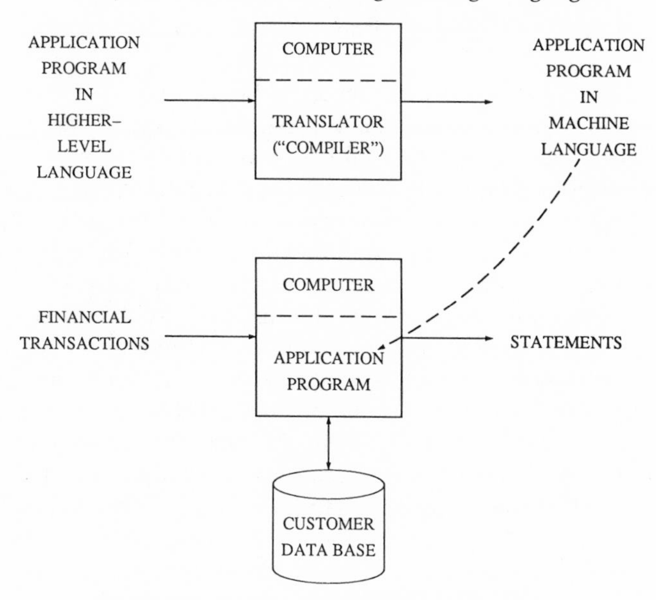

Figure 3-2. Steps in program translation.

cess shown in Figure 3-2. The first step is to translate the program expressed in a human-oriented language into an equivalent machine language program. Since this translation is a well-defined (but complex) procedure, it is thus suitable for the computer to do itself. Various types of translator programs exist (called **assemblers, compilers, interpreters, generators**, and similar names); they differ in their technical details, but perform basically the same function: converting a program for humans (called the **source** program) into a machine-language program for a given computer (called the **object** program). Under the direction of the translator program, the computer reads the source program as its input and produces the object program as its output.

The translated version of the program (i.e., the object program) is ready for execution. It is placed in memory, and control is then transferred to it. Now under the direction of the object program, the computer processes the program's inputs (e.g., check deposits and withdrawals) into outputs (e.g., bank statements).

In actual practice, a user need not be aware that a two-step process is required to execute a program. As far as the user is concerned, the computer could be interpreting a program directly without first going through the translation step. A user of a spreadsheet model expressed in Lotus 1-2-3, for example, logically can think of the computer as a Lotus 1-2-3 machine, even though hidden from view is a translation process.

For most commercial programming, the translation process is kept logically separate from execution of the object program. A source program goes through a period of development and testing, which requires translation for each test version. Once this has been completed, though, the final object program can be used for all subsequent executions without further translation (until the next program change).

Programming Languages

Choice of a Programming Language A wide variety of languages exist for most computers. Several criteria might be considered in choosing the language most appropriate for a given application:

- Ability of the language easily to express the kind of algorithms that arise in the application; in accounting applications, for example, sorting files and printing complex reports are common functions, while in engineering applications manipulating large arrays of numbers arises frequently.
- Availability of an **infrastructure** within the organization to support the language (e.g., trained programmers and related system software, such as **utility** programs to perform such common functions as sorting large data files).
- Hardware efficiency in translating the language's source program into the computer's object program.
- Relative hardware efficiency in the execution of the object program (compared with a machine language program written by an expert programmer, for example).

- Ability of programs written in the language to interface with existing programs used by the organization.
- Ability of the language to run on the hardware available within the organization (now or in the future).

First and Second Language Generations A machine language program represents **first-generation** software technology. Early **second-generation** languages were logically similar to the machine languages they replaced, except that they substituted **symbolic** references to operations and data, rather than using the binary codes used in the actual machine language. For example, rather than the machine language instruction 11000101010010101110001000101101, a programmer might write the symbolic instruction ADD A, AMOUNT. The ADD A defines to the computer the operation to be performed (adding a number in memory to the value in Register A), and AMOUNT is the symbolic name used by the programmer to refer to a given piece of data (the amount of a sales transaction, say).

In generating an object program instruction, the translator—typically called an *assembler* in the case of a second-generation translator—substitutes the binary equivalent of the ADD A operation, which it maintains in a table within the program. For the data item AMOUNT, the translator builds a table specific to a particular program that establishes a consistent (but arbitrary) one-to-one relationship between a memory address and the symbolic name of the data item stored at that location. Thus, for example, the translator might assign the AMOUNT data to the memory location having the address 1110001000101101. The fact that this relationship is completely arbitrary is immaterial as long as the translator maintains a consistent and unique relationship for all data references throughout the entire program.

Although allowing the programmer to refer to operations and data through symbolic references offers an important improvement over pure machine language programming, the approach still suffers from two crippling flaws. For one thing, it does nothing to reduce the number of instructions in a program, because for each symbolic instruction the translator (usually) generates one machine language instruction. A more serious flaw, though, is that this kind of language is not a good problem-solving tool. It is tied directly to the way the hardware instructions were designed, which bears little resemblance to the way people think.

Despite these limitations, machine-oriented symbolic languages were widely used from the mid-1950s through the 1960s, and even now find limited use. The translation process is straightforward, well understood, and efficient from a hardware standpoint. With suitable skill on the part of the programmer, the resulting object program can be made highly efficient in its use of machine resources. In the days when hardware costs were dominant, preoccupation with the machine efficiency of the translator and object program was regarded as legitimate by many MIS managers. (The fact that it resulted in serious long-term inflexibilities and inefficiencies is another matter, to be discussed later).

These languages currently do not find much use among application programmers, and then generally only for the purpose of maintaining old programs. The languages are still used fairly widely in writing system software, where machine efficiency can still provide an important competitive edge (particularly in the case of microcomputer software such as a spreadsheet program), but even here their use is declining rapidly.

Third-Generation Languages **Third-generation** languages—or **high-level** languages, as they were called at the time they were first introduced—were designed to give the programmer a more effective problem-solving tool. The design of these languages was relatively unencumbered by the need to preserve hardware efficiency by maintaining a close relationship between the language and the machine's **architecture**. Third-generation languages took the first significant step toward achieving machine independence by incorporating considerable computer intelligence in the translator, thus separating human problem solving from the idiosyncrasies of the hardware.

Fortran, the first (and still) widely used third-generation language, provides a good example. The programmer—an engineer, say—might calculate the area of a circle by writing the Fortran statement AREA = PI*R**2 (where R**2 is the Fortran version of R^2). This statement, along with a lot of other statements constituting the program (for designing a storage tank, say), is typically entered into a computer through an interactive terminal. When directed to do so by the programmer, the computer translates the statements into the object program. Each Fortran statement usually results in the generation of several machine-language instructions. The above statement for calculating the area of a circle, for example, might require as many as four instructions: (1) to retrieve R from storage, (2) multiply it by itself, (3) multiply the result by π and (4) store the final result in AREA.

This algebraic expression is closely related to the way an engineer might define the calculation, independent of the computer. As such, it provides a tremendous step forward in giving users an effective tool more closely related to the way they actually think about specifying computational algorithms. Each statement in a third-generation language accomplishes considerably more work than a machine language instruction, and therefore *human* efficiency (as opposed to *machine* efficiency) is much higher.

Third-generation languages dominate current practice. The vast bulk of existing application programs are written in these languages, and even in the case of newly written programs they enjoy a major (but declining) share. Most data processing programs are written in COBOL, while Fortran is favored for scientific and engineering programs. Other languages of this technical generation—some developed considerably later than Fortran or COBOL—offer certain advantages. Among the more popular of these are BASIC, PL/1, Algol, Pascal, C, and Ada.

Despite their widespread use and considerable problem-solving power, third-generation languages have some serious limitations. Although a program written in one of these languages has many fewer statements than the number of machine-language instructions, big programs still consist of thousands of such statements. Each is costly to develop. Each also risks introducing a serious bug in the program. To reduce this risk, programs have to be tested and debugged through a laborious and exhaustive process, adding greatly to the cost of program development. It is estimated that a programmer produces an average of only a few dozen or so fully tested and debugged statements each day (depending, of course, on the skill of the programmer and the complexity of the task).

These are by no means the only problems. Because of its size and complexity, a program is difficult to change. The problem can be greatly reduced by use of a **modular** structure in which program segments are compartmentalized in a way that allows a change in one module without having unforeseen consequences in other parts of the program. Nevertheless, even with the best design methodology, maintaining and adapting existing programs is a monumental and growing chore. Most organizations spend two or three times as much effort in maintaining old programs as they do in developing new ones.

Nonprocedural Languages Third-generation languages suffer from the intrinsic limitations of a **procedural** specification of a task. They require the programmer to define a step-by-step process for achieving a desired end result. In contrast, a **nonprocedural** language permits the programmer to define the task directly in terms of the end result. In defining to a taxi driver the task of getting from JFK Airport to the Empire State Building, for example, one could give highly detailed procedural directions for each leg of the trip, or one could merely say "take me to the Empire State Building." The end result (if not the precise route) may be the same, but the effort required to describe the task is altogether different.

A nonprocedural specification clearly requires more knowledge on the part of the interpreter than does a procedural specification. In giving a nonprocedural task description to the taxi driver, for example, one assumes that the driver knows the location of the Empire State Building and the routes for getting there. Ultimately, any task is executed through a procedure, whether by a person or a computer. The only question is, who supplies the intelligence to translate a "*what*-to-do" specification into a "*how*-to-do-it" procedure?

It is reasonable, for example, for Mr. Scrooge to ask his intelligent clerk, Mr. Cratchit, to "list the names of all customers in Birmingham who purchased £1000 or more from us last year." Mr. Cratchit can then use his own knowledge of the information system, as well as the meaning of Scrooge's terms ("customer," "purchased," "last year," "more than," etc.) to translate this request into a procedure, such as the following:

1. Select the customer ledger from the ledger library.
2. Open the ledger to the first customer record.
3. If the customer does not come from Birmingham, go to Step 6.
4. If the customer purchased less than £1000 last year, go to Step 6.
5. Write the customer's name on the list of names.
6. Turn the ledger page to the next customer record.
7. If the end of the ledger has not been reached, go to Step 3.
8. Give the list to Mr. Scrooge and stop the procedure.

Clearly, it is far easier for Scrooge to specify his needs in nonprocedural terms, and rely on Cratchit's ability to interpret his request. The nonprocedural form is not only much shorter—and therefore easier to communicate and less error prone—but it also can be formulated without any detailed knowledge of Cratchit's data processing procedures. Scrooge presumably is only interested in the end result, the customer list, so why should he have to concern himself with processing procedures?

A nonprocedural computer language is built on this simple and fairly obvious principle. There are limits in applying this principle, however: translation of a nonprocedural task description into an equivalent procedure always requires knowledge of the subtasks that arise in a given class of problem. Designers of such a language thus face a difficult tradeoff between the generality of the language—the range of problems for which it can effectively be used—and the amount of knowledge that it can incorporate about the tasks for which it is intended.

Financial analysis, for example, frequently calls for the calculation of the net present value (NPV) of a stream of cash inflows and outflows. Accordingly, for the computer to respond successfully to the task description, "Calculate the NPV of the interest receipts from the Series 1998 bond" (suitably stated in the syntax of the language), the translator program must know a number of things about NPV and the bond portfolio. It must know about the structure of the database so that it can locate the appropriate cash flow data and the discount rate, and it must have a procedure for computing NPV. For similar reasons, a nonprocedural language intended for engineers must know about such arcane matters as solving differential equations and inverting matrices; one intended for lawyers must know about searching through legal citations for specific case references.

If the translator program does not have sufficient knowledge to interpret a nonprocedural task definition, the human programmer must supply the knowledge through a procedural specification expressed in a more primitive form. For example, the specification for determining a NPV might call for the calculation of each period's cash flow, followed by the multiplication of this value by the appropriate discount factor and adding the resulting product to the cumulative NPV. Having to program at this level of procedural detail obviously shifts much of the burden, and much of the knowledge base, from the automatic translator to the programmer.

No language can know everything about everything, and so the tradeoffs generally favor having a given language deal knowledgeably with a fairly narrow set of problems. As a result, we will continue to see a proliferation of special-purpose languages that are very good at the limited things they do. The Tower of Babel in computer languages will not crumble away.

Fourth-Generation Languages We are now at a fairly early stage of the widespread use of **fourth-generation** languages (**4GLs**). Although there is no clear consensus as to the definition of a 4GL, most of the languages labeled as fourth generation have many of the following characteristics:

- Nonprocedural task specification, taking the form primarily of an inquiry and report generation language (allowing, for example, inquiries quite similar to Scrooge's inquiry about Birmingham customers).

- Close integration with a database management system, allowing data to be selected from a database and manipulated by the 4GL (summarized, formatted, displayed, etc.).
- Strong emphasis on ease of learning and ease of use (which are not the same thing)—important aspects of **user friendliness** designed to permit the nontechnical user to formulate his or her own information needs directly in the 4GL itself.
- Interactive orientation, designed specifically to take advantage of a close interaction between the user and the computer (e.g., facilitating quick trial-and-error calculations so that little penalty is experienced in time, cost, or risk if the user commits an "error").
- Visual orientation—for example, use of stylized symbols, or **icons**, to represent operations in the system (such as a trash can to signify the operation of deleting a stored document).
- Use of **windows**—flexible partitions of the screen, which can be created, moved, changed in size, and deleted by the user; each window can be devoted to its own independent display, with coordination across windows (such as two different documents displayed in separate windows, with portions of each combined into a third document displayed in a third window).
- Non-keyboard orientation—for example, use of a pointing device (such as a **mouse** or even one's finger) to identify a character or location on the CRT screen.
- Inclusion of a set of development tools designed to increase the productivity of the programmer (e.g., a **screen formatter** for easily defining the contents, format, and editing specifications of a CRT screen used for data entry).
- Incorporation of a **natural language processor** as the human interface to the system, permitting the computer to respond automatically to a user's information needs expressed in a relatively unconstrained version of a natural language like English or French (*exactly* as stated by Scrooge, for example); in fact, natural language processing is one of the defining characteristics of the so-called **fifth generation** languages that are conjectured to be available in the 1990s.

Not every language touted as a 4GL has all these characteristics, but the trend is certainly in the direction of increasing the number and power of these features.

A wide variety of 4GLs are now on the market, and cover the full range from micros to mainframes. They are available from both hardware companies and third-party software vendors. A microcomputer product may sell for a few hundred dollars (depending on one's definition of a 4GL), while a mainframe product can sell for well above $250,000.

What we now consider to be 4GLs first appeared on the scene during the 1970s. An early example is Ramis, followed somewhat later by Focus; both are still widely used. Though they began as fairly specialized languages for query processing and report preparation, their capabilities have subsequently been expanded to include more generalized data manipulation and database management functions. Another avenue of development has been the addition of a query language and other development tools to a general database management system; Cullinet's IDMS and Software AG's Adabas, with their evolving sets of associated products, have followed this path, for example. Thus, the history of development of 4GLs has been to start with a query and reporting language and then to expand downward to database management, or, on the contrary, to begin with a DBMS and expand upward to provide enhanced 4GL tools.

A substantial payoff can be gained by moving to a 4GL as the preferred language for MIS development. In fact, the defining characteristic of a 4GL, according to some experts, is its ability to achieve a tenfold improvement in programmer productivity. Most current experience falls short of this goal, but the potential certainly exists for dramatic gains. With such improvements come significant *qualitative* changes in the way systems are implemented:

- The emphasis shifts to how users can better define their information needs, away from a preoccupation with the technical engineering aspects of an MIS; this shift is possible because it becomes relatively cheap, fast, and low-risk to implement a functioning system once a working specification has been fashioned.
- Since a high-productivity development tool lowers the cost of being "wrong" in an initial statement of requirements, it becomes entirely practical to use trial-and-error **prototype** systems, "throw-away code," and other quasi-experimental meth-

ods aimed at promoting an evolutionary process of organizational learning and systems adaptation.
- The relative ease of learning and using 4GLs permits the widespread diffusion of the skills necessary for users to interact directly with the MIS, without the aid of technically trained intermediaries.

It is difficult to overstate the importance of these effects in making the MIS more responsive to user needs and better positioned to serve the strategic goals of the organization. Unfortunately, however, the transition from third-to fourth-generation programming languages is neither easy nor quick. Although some organizations have already made substantial progress along the transition path, most have not.

Large organizations, in particular, have an enormous investment in application software and in the skills of their MIS staff. Interfacing 4GL-based new systems with the existing programs is usually not easy. Furthermore, most organizations will find a continuing need to develop some of their new applications using the more general or machine-efficient third-generation languages, especially in the case of high-volume transaction processing systems that push against the capacity limitations of the fastest available machines. As a consequence, we are likely to see programs written in COBOL, Fortran, and their ilk surviving well into the 21st century.

Guidelines Concerning Hardware and Software

The bewildering array of hardware and software offered in the marketplace often makes it difficult to select the appropriate configuration for a given application. However, there are some insights and rules of thumb that a senior manager can use to understand these matters:

- Any design or hardware/software selection decision motivated primarily by a desire to gain hardware economies is usually wrong.

- As a corollary to the above guideline, an MIS can incorporate computer processing power at an acceptable hardware cost wherever functional needs require such intelligence.
- Software, not hardware, should drive almost all MIS design decisions.
- Software generally has a much longer (albeit evolving) life than the hardware on which it runs, and the database has greater longevity than either of them.
- The choice of a hardware architecture is much more important and enduring than the choice of a particular piece of hardware.
- One should not migrate the MIS from one hardware architecture to another without some very good reasons.
- Connecting one vendor's hardware or software to another's is generally difficult unless the components were designed by a single vendor explicitly to fit together homogeneously (or proluced by a **plug-compatible** vendor to adhere to a well-defined standard).
- Connecting hardware or software from different vendors is almost always technically *feasible* (within wide limits, at least); whether or not it is *desirable* depends on the tradeoffs between the benefits of heterogeneity—primarily the ability to select the best combination of products from a variety of vendors—versus its added complexity.
- A 4GL should be used for software development unless a very good argument exists for using a conventional language such as COBOL or Fortran.
- With the widespread use of high-productivity software development tools, implementation responsibility shifts more strongly from programmers to the users who define the requirements of the system.
- Information requirements can never be defined adequately the first time (or perhaps ever), and so the system must be capable of undergoing continual adaptation and growth.

Further Readings

This material is covered in the general MIS textbooks listed as additional reading for Chapters 1 and 2 (for example, Davis and Olson, Lucas, and Senn).

❖ 4 ❖

Transaction Processing

Critical Nature of Transaction Processing

What Is Transaction Processing?

Transaction processing provides information-processing support for operations throughout the organization. Increasingly, transaction processing is moving toward interactive support of virtually all operational activities, reaching to all corners of the organization, and even beyond, through the telecommunications network. It provides the primary means of updating the database to keep it current with real-world events.

In a large transaction-processing system, such as an airline reservation system, several hundred transactions per second might be processed during peak periods. The technology pioneered by leading companies is now quite mature. The hardware is cheap enough, and the software is widely enough available, that virtually any organization can seriously consider the implementation of an interactive transaction processing system.

A **transaction** is an event of significance to the MIS. The event typically requires that some sort of activity take place within the organization to accomplish the purpose of the transaction, such as shipping a product in response to a customer's order. The database is updated to reflect the event, perhaps in a number of different records (e.g., a sales order may result in the updating of accounting records, production schedules, and purchase orders).

In general, then, a transaction gives rise to a data input into the system; this input is processed by the system, and one or more outputs are generated. The following are three typical examples of such processing:

1. A hospital laboratory completes a test for a patient; data entered about this event cause the updating of the patient's medical and accounting records, as well as a report to the attending physician.
2. Payroll data enter the system, describing an employee's hours of work, the allocation of the hours among jobs, changes in job classification, etc.; the payroll program then generates such outputs as paychecks or direct bank deposits, labor cost distribution reports, and updated payroll records.
3. A bank customer's check is presented for payment; the system reduces her bank balance by the amount of the check, and retains data needed to prepare her next statement.

Why Is Transaction Processing So Important?

Transaction processing is a critical part of the MIS. For one thing, it generally accounts for most of the cost of operating the system. Although the other major part of the MIS—the **decision support system** aimed at assisting decision makers throughout the organization—is also vital to the success of the enterprise, it normally requires a fairly small share of the computing resources.

In a comprehensive MIS, most of the organization's day-to-day activities are handled through the transaction processing system. If a problem arises—an item gets shipped to the wrong address, say, or a customer invoice is incorrect—the fault generally lies with the transaction-processing system. Improvements designed to give the organization a strategic advantage through better service, higher quality products, or lower costs often come through enhancements in the transaction-processing system.

As the MIS becomes more involved with operational matters, the organization's efficiency and effectiveness become more dependent on the intelligence built into the transaction processing system. Consider, for example, the case of an order entry system for a wholesale distributor. When the system encounters a stockout condition for one or more items on a customer order, it has the following options:

- The order can be canceled entirely.
- The entire order can be backordered—that is, retained for later shipment when the missing items become available.
- Items that are in stock can be shipped at once, with the missing items either canceled or backordered.
- If a missing item is currently on order, the system can determine when the item can be expected back in stock (and possibly expedite the current replenishment order from the supplier).
- If a missing item is not currently on order, a replenishment order can be generated, with appropriate expediting.
- If the order entry system is **interactive**, as it increasingly is, the parties to the transaction can participate in the actions taken by the system—deciding, for example, whether an out-of-stock item should be backordered or canceled depending on the expected delivery date predicted by the system.

Critical Points About Transaction Processing

The example of handling a stockout illustrates some exceedingly important general points about the transaction processing system.

First, it should be apparent that the way the transaction-processing system handles backorders has a critical impact on customer's attitude toward the company, as well as on operating efficiencies within the company. The system treats all customer orders in the same consistent fashion (although, of course, an intelligent system will take account of each customer's idiosyncratic requirements). It follows, therefore, that the cumulative effects of the order entry system are important, pervasive, and enduring—and therefore strategic. Any serious search for excellence by top management must surely include concern with the organization's transaction-processing system and its continual adaptation to meet changing needs.

Second, the transaction-processing system incorporates extensive knowledge, experience, and policies—computer "intelligence"—that define how the organization deals with operational activities. These are largely matters of *management* responsibility and expertise, not *technical* issues. Employees who have a stake in the behavior of the system, and have expertise to contribute (including, it is important to note, workers at the bench level directly involved with operations), should therefore play a role in establishing the system's specifications. Such issues should certainly not be left solely to the discretion of technical staff personnel, no matter how competent they might be.

Third, the transaction-processing system becomes the primary vehicle for getting things done in a comprehensive MIS. Efforts to improve operations and services to customers—in handling backorders, as just one example—should therefore come mainly through improvements to the transaction-processing system. To incorporate the organization's collective knowledge and judgments about operational matters, the system must be kept flexible so that it can adapt to organizational learning.

Fourth, managers should regard the transaction-processing system as a fundamental mechanism for making long-lasting adaptive improvements in the behavior of the organization. Thus, if an operational problem arises, the system should be modified so that the problem will not occur again. For workers and organizations used to a firefighting mode of problem solving, where each new crisis is handled in an ad hoc fashion, attacking problems through adaptation of their information system generally requires a difficult change in culture and attitude.

Finally, the transaction-processing system provides the primary entry point for data about the real world. For example, the order processing, purchasing, and accounts receivable subsystems for a wholesale distributor provide a great deal of data about the basic merchandising cycle within the firm—who buys what, when, how many, and at what price and credit terms; the payment and merchandise return history of each customer; the delivery and quality performance of each supplier; and the performance of each buyer and salesperson. As the formal transaction system pervades the organization, the database grows in importance as a repository of extremely valuable operational information.

Design of the Transaction-Processing System

It will be worthwhile at this point to consider design issues with somewhat more focus than was possible in Chapter 2. In illustrating the issues, we will stay with the previous example of an order entry system for a wholesale distributor.

Data Entry

The basic transaction cycle for the business begins with the receipt of a customer order, which may enter the system in a number of different ways:

1. Through the regular mail system, with the order prepared by the customer or by a salesperson of the wholesale firm.
2. Through electronic mail, similar in principle to regular mail but without the delay and uncertainty of the postal system.
3. By telephone, with an order entry clerk using an on-line terminal to enter the sales information while the customer remains on the telephone to answer questions and make decisions about the order.
4. Through an in-house **point-of-sale (POS) terminal.**
5. By remote data terminal, operated by the wholesale vendor's

salesperson using a portable terminal, or by the customer's own personnel.

6. By touch-tone telephone, allowing the customer to punch keys on the instrument to identify the customer, items ordered, and quantities; instructions and feedback information to confirm the keyed values are then provided automatically by the computer through the use of a **voice synthesizer** that converts digital data into high-quality spoken form.

Once entered, the data are then edited for accuracy. One level of editing looks at the transaction fields to assess their accuracy and plausibility. For example, the catalogue number of an item might be edited to determine if it has the correct number of characters and whether the characters are of the correct mode (alphabetic, numeric, or mixed). More thorough editing requires access to the database to judge the correctness of the transaction data. An entered product number, for example, might have the right length and structure, but may have no product in the database associated with it. Errors of this sort can only be detected by checking the inventory file.

Eventually a transaction gets edited against the database; the only question is whether this happens at the point of data entry. If done interactively as the transaction is initially entered, the data entry terminal must be connected to an on-line database while the editing takes place. Thus, transactions that enter from points scattered throughout the organization require either the maintenance of an up-to-date database at each location, or a continuous telecommunication connection from each transaction site to a central database.

A continuous communication link is fairly expensive. Communication cost can generally be reduced if transactions are handled by a stand-alone local computer with its own database. Any updating of the central database can be done by accumulating transactions at each local processor and then transmitting the data periodically to the central computer site (possibly only in summary form). As a further source of saving, the system can transmit the data during an off-peak time period when common carrier rates are low or the internal network offers idle capacity.

This approach does not provide a central authoritative source of up-to-the-minute information throughout the day. Applications that require such timeliness—an airline reservation system is an obvious case—can be accommodated, at a cost, by continuous communications. For many applications, however, interactive data entry coupled with daily batch updating of a central database provides a good balance between economy and timeliness.

Content of the Database

The content of the database is increasingly being recognized as one of the really strategic issues in the management of information systems. Individual applications are ephemeral: they can come and go, and be modified and extended, with relative ease. The database, too, is usually modified extensively throughout the working day, but its general content and structure tend to be more stable than application programs. As a result, the design of a transaction processing system should focus primarily on the database and only secondarily on the individual applications. If the database design is sound, the applications are generally relatively easy to handle.

Figure 4-1 gives a simplified database structure for the wholesale distributor. The information shown is organized into five different **record** types, describing customers, inventory items, suppliers, customer orders, and purchase orders.

As shown in the figure, each record consists of individual pieces of data, called **fields** (or **data elements**). A customer record, for example, contains fields for the customer number, name, address, and the like. A field generally constitutes the lowest-level component or "atom" in the database.

Designers typically need to engage in a certain amount of speculation in judging how the system will be used. For example, if the customer's address is used only in preparing mailing labels, then a single address field can be defined to include the street number, street name, city name, state, and ZIP code (with the appropriate line spacing). If, however, a

CUSTOMER RECORD	*INVENTORY RECORD*
CUSTOMER NUMBER NAME ADDRESS CLASSIFICATION ACCOUNTS RECEIVABLE STATUS PROFITABILITY INDEX HISTORICAL DATA	ITEM NUMBER DESCRIPTION CLASSIFICATION SUBSTITUTES ORDER POINT ORDER QUANTITY COST PRICE LEAD TIME SUPPLIER NUMBER(S) ON-HAND BALANCE ON-ORDER BALANCE HISTORICAL DATA
SUPPLIER RECORD	CUSTOMER ORDER
SUPPLIER NUMBER NAME ADDRESS QUALITY RATING ACCOUNTS PAYABLE STATUS HISTORICAL DATA	CUSTOMER ORDER NO. CUSTOMER NUMBER ORDER DATE SHIP DATE LINE-ITEMS ITEM NUMBER QUANTITY
PURCHASE ORDER	
PURCHASE ORDER NUMBER SUPPLIER NUMBER P.O. DATE SHIP DATE LINE-ITEMS ITEM NUMBER QUANTITY	

Figure 4-1. Simplified database organization.

finer definition of address is required, then each of these parts of the address (e.g., the ZIP code) might be defined as a separate field (so the addresses can be sorted by ZIP code, say). Although it would certainly be possible for a skilled person to program the computer to do this if the ZIP code were a part of a single address field, it is far easier (especially for a nontechnical user) if the ZIP code is defined as a separate field (e.g., by using a 4GL that permits a statement such as PRINT NAME, STREET, ADDRESS, CITY, STATE, ZIP BY ZIP).

A given *occurrence* of a record pertains to a particular entity, such as a given customer or inventory item. For example, the customer record for S & P Sales might contain the data shown in Figure 4-2.

The database consists of records of various types. Each record included in the database describes an entity of (presumed)

CUSTOMER NO.:	3813
CUSTOMER NAME:	S & P SALES
ADDRESS:	1385 GLOVER AVENUE PRINCETON, NJ 08540
CLASSIFICATION:	K-B-A
A/R STATUS:	BALANCE $4,540 CREDIT LIMIT $10,000
PROFITABILITY:	$8,720
HISTORICAL DATA:	FIRM ESTABLISHED 1949 FIRST PURCHASE AUG 1972 NO PAYMENT PROBLEMS

Figure 4-2. An occurrence of a customer record.

relevance to an application or decision process. Each occurrence of a record is first created through a record-initiation transaction, and it is kept current through the updating that occurs when subsequent transactions are processed by the system. For example, the record for S & P Sales was created in August 1972 at the time of the company's first purchase from the wholesaler; a field such as the current credit balance is revised as events take place (purchases, payments, returns, credits, and so forth).

Database Processing

Frequent access to a database is a universal requirement of a management information system. This applies to all parts of the system, but it is especially critical in a transaction-processing system because of the high volume of activity and the need for quick access to operational data.

Providing efficient and effective access to the database is one of the most challenging tasks facing MIS designers. The database may be huge, consisting of billions or even trillions of bytes. When a transaction enters the system, the computer must be able to match the transaction data with the corresponding database data to complete the processing. A sales order, for example, must be matched with data in the customer's account (current credit balance, credit limit, etc.), with the inventory records for each item ordered, and perhaps with suppliers' records if the order requires the purchase of replenishment items. When the appropriate data from the database have been read into main memory, the computer can generate such outputs as shipping labels and invoices. The computer concludes the processing by updating the database to reflect the actions taken.

The two basic ways of matching transaction data with the corresponding data from the database are **sequential processing** and **direct access processing** (also called **on-line processing** or **random processing**; although technical distinctions exist among these terms, they are generally used more or less synonymously).

Sequential Processing Consider our example of processing a customer order. This typically calls for retrieval of the customer's record from the database to check the customer's credit standing. With sequential processing, retrieval of the data is achieved through a sequential scan of customer records. To make this acceptably efficient, all such records are stored together in the customer **file**. The computer reads one record after another from the file until the desired record is found—that is, until the computer encounters a customer record that has the same customer **identifier** as the transaction record it is currently attempting to process. The identifier may be the customer's name or, more likely, a customer number assigned for the purpose of providing unique and condensed identification.

It is important to note that the same identifier must be used for the transaction record and the customer record. The person entering the transaction must thus be able to associate the customer with its identifier. If the identifier is not known, it can be determined by looking it up in a table in which the number is associated with some other company identification, such as its name and/or address.

Suppose that the customer file is stored sequentially on disk. Suppose further that it takes one minute to read the entire file (which is a representative time to read a sequential file). With good luck, the record being sought will be found near the beginning of the tape, in which case the search will take only a few seconds. With bad luck, the record will be near the end, taking one minute to locate. Records to be retrieved from the file will ordinarily be distributed more or less uniformly along the length of the tape, so on average half the tape must be read before the target record is found. The average access time to retrieve a record will therefore be about 30 seconds under the hypothesized conditions.

To take this amount of time to process a single transaction would obviously be absurd. We have a fairly simple remedy, however: rather than searching the customer file for a single record, a **batch** of many transactions is processed together in a way that requires only one complete scan (or *pass*) of the tape. If, say, 1000 sales transactions are processed in the batch, and the time to read the tape is still only one minute, or 60 seconds, then the access time per transaction is reduced to 60/1000 = .06 seconds (60 milliseconds)—still not exactly a breakneck speed, but certainly a vast improvement over the single-transaction approach.

The trick in gaining this advantage is to arrange things so that all transactions can be processed in a single pass of the file. This is done by sorting both the sales transaction records and the customer records into the same sequence prior to scanning the file. In our example, both collections of records—the transaction file and the customer file—would be sorted on the customer number, in ascending sequence, say. (The customer file stays in this same sequence, so it does not have to be sorted each time it is processed.) Both files are then read in this sequence, in synchronized fashion, with matches between the two files being found along the way. As each match is found, the computer has all the data necessary to process the matched transaction. When the computer gets to the end of the customer file, all transactions will have been processed.

Several very important consequences stem from the batch approach to transaction processing. The most obvious one is the processing delay introduced by the batching interval. The method is only efficient if many transactions are processed together in a single batch, which is achieved primarily by accumulating transactions over an extended period of time (generally a day or more). A transaction entering just after the cutoff point for the preceding batch has to wait during the full batch accumulation cycle before it gets processed and the database gets updated.

A less obvious but more fundamental problem also exists: relationships between records—pertaining to an inventory item and its supplier, say—are very difficult to handle with sequential processing. This is so because the computer processes only one database file at a time. In fact, only a single record of the current file is accessible at any given moment. Once a record has been updated, it is written back into external storage and is no longer retained within the computer's main memory (since the main memory is not large enough to store all processed records). Thus, there is no way that the computer can gain access to another record, such as an inventory or supplier record, while processing a given customer record. The only thing that can be done is to retrieve the other records when next processing the file on which they are stored. Bringing all the pieces together again to complete the processing raises a number of additional complications. Under this scheme, retrieving information from multiple records is so difficult that it is avoided whenever possible.

Despite these limitations, batch processing still continues to be a commonly used form of transaction processing. This is due partly to the fact that existing applications were installed years ago when batch processing with low-cost magnetic tape storage offered distinct cost advantages compared to alternative methods requiring relatively expensive on-line storage. Many of these obsolete applications are now being converted to a more modern design. Even with the best contemporary design, though, batch processing still has an important role. Certain types of transactions, such as payroll and billing, are intrinsically periodic in nature and therefore lend themselves well to batch processing.

Direct Access Processing As we have seen in Chapter 2, a direct access storage device such as a magnetic disk drive is able to access a specified storage location directly without having to access all intervening locations. Each such location—often termed a **sector**—has a unique physical **address** that allows the computer to refer to it when it wants to read from or write into that location. For simplicity's sake, we can consider that each database record is stored in an addressable disk sector. (In fact, one or more records might be stored in a single sector, while a long record might require several sectors.)

With the database stored on disk, processing a customer order is greatly simplified. All we need is the disk address of the customer record to determine if the customer passes the credit screening. Using this address, the computer can directly access the record and read it into memory for processing. The updated record can subsequently be written back into the same disk location.

We have a problem, though, in determining the address of the record. The solution is not necessarily obvious, but it turns out to be fairly straightforward.

It should first be noted that the transaction record itself would rarely contain the physical address of the corresponding database records. The database records are scattered around on the disk, and may even be moved from time to time to take advantage of new disk technology or changes in the database contents and structure. The last thing we would want to do is to freeze the physical organization of the database to provide stable disk addresses to users of the system. Even more damning than its technical rigidity, such an approach violates a basic tenet of good systems design: avoid, if at all possible, inflicting pain, suffering, and inconvenience on users simply to make life easier for system designers and programmers. Fortunately, there are perfectly practical ways to associate an internal disk address with information known to users—in this case, the customer identifier.

Conceptually the simplest way to make such an association is by having the computer look it up in a table or index maintained by the system. Suppose that we want to retrieve the customer record for S & P Sales, Customer Number 3813. This record is stored, let's say, at disk address 70335. In the index we simply need an entry that provides such an association: 3813 → 70335. Each such entry is maintained automatically by the system software, with no need whatever for the user to be concerned with the disk location of a desired record.

This simple description omits a number of details. The important point to note, though, is that straightforward methods exist for relating a record's external identifier with its internal address. Once the internal address is found, the computer can retrieve the desired record by accessing it directly from disk storage. Although the underlying process may be quite complicated, the system software shields the user from many of the details.

Direct access processing provides the obvious advantage of quick access to the database. Of even greater importance, however, is the fact that with direct access storage the computer can access all of the on-line database. This permits it to handle all ramifications of a transaction. For example, after checking a customer's credit rating, the computer can immediately look at the inventory balance of each item included on the sales order. If an item is out of stock, the computer can retrieve data about the item's supplier to determine the delivery date of a replenishment order.

Quick access to related records is virtually impossible with conventional sequential processing. This helps to account for the rapid shift from batch processing with magnetic tape storage to direct processing with magnetic disk storage. Although batch processing will continue to play an important role in information systems, the trend is definitely toward the use of interactive workstations linked to an on-line database that supports direct transaction processing. This same database can also be used for batch processing applications requiring data that enter through the on-line system. The use of a common on-line database to support both interactive and batch applications is, in fact, rapidly becoming the preferred choice for most management information systems.

Further Readings

This material is covered in the general MIS textbooks listed as additional reading for Chapters 1 and 2 (for example, Davis and Olson, Lucas, and Senn).

❖ 5 ❖

Decision Support Systems

What Is a Decision Support System?

A **decision support system**, or **DSS**, is designed to aid human decision making and provide productivity tools for knowledge workers. Transaction processing, in contrast, deals with routine operational matters. The DSS can thus be defined as all of the MIS except for transaction processing.

Although it is useful to draw a distinction between transaction processing and decision support systems, in practice the boundary between them may be quite fuzzy. Increasingly, we are likely to see each manager and professional staff member, along with clerical and operational personnel, equipped with a multifunction workstation linked to a corporate network. The same system that serves the user's need for decision-oriented information can also be used to enter transactions, selectively retrieve information from a transaction-fed database, or obtain a variety of office functions such as word processing and electronic mail.

Most current management information systems devote the bulk of their attention and resources to transaction processing, with only incidental aid provided to decision makers. This lopsided allocation of resources is likely to continue, due to the heavy demands placed on the system by the profusion of interactive terminals being installed to support transaction processing. Nevertheless, increased attention is being focused on making the MIS more responsive to the information needs of decision makers at the operational, tactical, and strategic levels of the organization.

A number of factors account for the current high level of interest in DSS:

- A growing sophistication about computer-based systems on the part of managers and professional workers.
- A growing recognition by managers that the MIS can and should be designed to meet their need for decision-oriented information.
- Technical developments—primarily low-cost hardware and user-friendly software—that make computing services much more accessible than before.

Types of Decision Support Systems

A wide variety of DSS have been implemented. They differ greatly in the functions they provide, their complexity, and the time and cost necessary to develop them. A few examples might serve to illustrate this variety:

An automobile dealer uses a spreadsheet program on a personal computer to compute the financial terms for auto loans.

A staff assistant to the controller uses an inquiry language to access the corporate database and prepare ad hoc reports analyzing the distribution of finished goods inventory by product and region.

The Treasurer's office uses a sophisticated model to predict cash receipts and expenditures to reduce unproductive cash float in the company's bank accounts.

The Vice President of Finance for a modest-sized retailer develops a spreadsheet model that generates predicted *pro forma* financial statements under various assumptions regarding the level of sales, gross profit margins, and operating costs.

The Distribution Department of a manufacturing company uses a model to plan the truck routing that minimizes the cost of delivering products from production facilities to customers.

The Marketing Department uses a model to determine the allocation of advertising funds among media to minimize the cost of achieving a given level of exposure to persons having specified demographic characteristics.

These examples illustrate a number of important points about DSS:

- A DSS may be used to generate straightforward summary reports (as in the case of the inventory analysis reports) or to execute complex mathematical models (as in the truck routing example).
- A DSS may be executed on a personal computer using a simple spreadsheet language (for the auto dealer) or on the company's central mainframe computer (to perform the cash management function for the Treasurer's office).
- The DSS may be developed and operated by the direct user (the VP of Finance of the retail company), by a support person working for the decision maker (the staff assistant to the Controller), or by the technical staff (in the case of the truck routing application).
- The database for the DSS may be self-contained (the auto loan program), integrated with the corporate database (to analyze actual inventory balances), or linked with a proprietary external database (demographic data coming from surveys of the readers and viewers of alternative advertising media).
- The benefits of a DSS may be expressed in quite tangible monetary terms (savings in truck delivery costs) or only in intangible nonmonetary terms (better financial planning for the retailer).

Decision Aids Provided by a DSS

A decision support system provides computer-based assistance to a human decision maker. This offers the possibility of combining the best capabilities of both humans and computers. A human has an astonishing ability to recognize relevant patterns among many factors involved in a decision, recall from memory relevant information on the basis of obscure and incomplete associations, and exercise subtle judgments. A computer, for its part, is obviously much faster and more accurate than a human in handling massive quantities of data. The goal of a DSS is to supplement the decision powers of the human with the data manipulation capabilities of the computer.

A designer of a DSS must choose the tasks to allocate to the computer; the remaining tasks are then left for the human decision maker. The appropriate degree of aid for a given decision process can vary from none at all to complete computer-determined decisions with only a monitoring role played by the human.

No DSS Aid

Many important decisions are, of course, made without any significant role played by the computer. To some extent this is a transient condition that will change as computer-based decision aids become more widely used within organizations. Nevertheless, there will always remain problems for which a DSS is not deemed appropriate because of their nonquantitative nature or because they are not well enough understood for formalization in a DSS.

Most existing DSS deal with strictly quantitative variables, data, and relationships. A cash flow model, for example, might handle cash receipts and expenditures in relation to a given production and sales pattern. The outcome variables (e.g., net cash inflow or outflow during a given month), the input variables (e.g., production and sales for each month), and the relationships among the variables (e.g., the time lag between a sale and the corresponding cash receipt) can all be expressed in well-defined quantitative terms.

Decision aids have not been widely applied for decision making involving multiple, nonquantitative, and conflicting goals.

In passing tax legislation, for example, politicians have to deal with such goals as "equity" and "economic efficiency" in a little-understood environment of different and often competing political philosophies, constituencies, regions, generations, industries, and income strata. A problem such as this generally defies any consensus as to what variables are important, how they can be measured, and the relationships among them. Under these circumstances, a DSS may not have much impact on the decision process.

Even if a DSS could, from a technical standpoint, provide aid for a given decision, its expected payoff might be too low to justify the cost of implementation. The payoff depends on the stakes involved in a decision, the likely improvement that can be achieved with the aid of a DSS, and the frequency with which the decision is made (e.g., a one-shot bank loan versus a recurring series of loans). The cost of development depends on such factors as the complexity of the problem, the difficulty of collecting data, and human interface requirements of the DSS (e.g., graphics versus tabular display of outputs).

New developments have substantially expanded the variety of problems for which a DSS can make a useful and economically justified contribution. A number of advances have contributed to this result:

- The drastically lower cost of hardware has made personal computers and interactive terminals readily accessible to decision makers.
- Easy-to-learn nonprocedural languages permit user organizations to develop their own DSS without having to bear the incremental cost (and possibly the red tape or even hindrance) of the central MIS staff.
- Advances in text processing software permit the manipulation and analysis of qualitative information encountered in nontraditional DSS applications.
- The relative ease and low cost of developing a DSS often make it attractive to provide partial quantitative aid for an ill-structured and basically nonquantitative problem (such as analyzing the tax burden imposed on different income groups by alternative tax proposals).
- Analytical methods and software products have been devel-

oped to assist decision makers in assigning subjective weights to multiple objectives to find an alternative that provides an acceptable balance among conflicting goals.

Selective Retrieval of Information

One of the most useful aids that can be provided to decision makers is easy access to selective information in the form of summary reports, responses to **ad hoc queries**, **exception reports**, and graphical displays. Most of the needed information can be derived from raw data in the organization's database; additional information might also be obtained from public sources, such as a proprietary **econometric database.**

A system that generates such distilled information is termed a **data-oriented** DSS. Compared to a more sophisticated **model-oriented** DSS, which relies on a model of some sort to predict the consequences of alternative courses of action, a data-oriented DSS provides a relatively low degree of aid to the human decision maker. The human is left with the task of generating and assessing alternatives, and choosing an alternative from among those considered. In the course of these subjective evaluations, the decision maker may probe the DSS repeatedly for additional information in trial-and-error fashion.

Despite the fact that a data-oriented DSS offers a relatively low level of support, the information obtained from such a DSS may have great value. A frequent, if not universal, complaint of the users of most existing MIS is that the system does not provide useful information for decision making. The system may generate voluminous reports, but often they are not organized and selected in a way that facilitates decision making. Users of such reports are left with the burden of manually selecting the information they need.

The effort in selecting useful information from voluminous reports may well exceed the information's value. As a consequence, users may be denied access to valuable information because it is buried in scattered and undigested form throughout the reports or in a huge database. Such information might just as well not exist, for all the good it does. Unless steps are taken to remedy the situation, the problem of data pollution can only get worse as the transaction-processing component of the MIS becomes more pervasive and the size of the database continues to grow at an explosive rate.

A data-oriented DSS is designed specifically to deal with this issue. A good DSS serves as an effective filter in screening out the vast quantity of irrelevant data, presenting users with selected information for decision making. By providing an easy means for a user to "play around with the data," the DSS allows the decision maker to gain greater understanding of a problem. Eventually the problem may be well enough understood to permit the development of a formal decision model that gives a higher degree of decision aid.

(It might be noted parenthetically that a distinction is often made between the terms **data** and **information**. The former is raw and undigested, generally entering the MIS as transaction data. Information, on the other hand, is derived from data with the intention of making it available for decision making. We will loosely adhere to this usage, although in practice the distinction is often not very clear.)

Decision Models

A decision **model** gives an abstract representation of a real-world situation. The outcome predicted by the model depends on both the choice of controllable actions as well as the forecasted value of uncontrollabe variables. This can be described succinctly by the following equation:

Outcome Variables = Model(decision variables,
uncontrollable variables)

Using the model's predictions of the outcomes for alternative decision values, the decision maker can choose a course of action that appears to meet his or her goals.

In a production planning model, for example, the decision variables might be the aggregate regular time and overtime production levels in labor hours over the next twelve months; the uncontrollable variables might be the forecasted sales and the cost factors involved in the production process. Based on the chosen level of production and the forecasted values of the uncontrollable variables, the model calculates such outcome variables as the expected manufacturing cost and the level of inventory over the planning period. By examining the predicted consequences of alternative production levels, the decision maker can search for the production plan that minimizes manufacturing cost subject to inventory constraints (e.g., a maximum or minimum limit on the permitted level of inventory). Whether or not the *actual* consequences of an implemented plan match the predicted outcomes depends on the accuracy of the model and forecasts of uncontrollable variables.

The issue of controllable versus uncontrollable variables deserves some amplification. Most variables are controllable by *someone*. For a variable to be considered uncontrollable in a given situation, the decision maker for whom the model is developed would normally have an insignificant effect on the value of the variable. For example, a manager of inventory control would typically have relatively little influence on sales demand, and so such demand could legitimately be considered uncontrollable in an inventory model. A media selection model for the marketing manager, however, would certainly not treat demand as uncontrollable.

As a practical matter, the distinction between controllable and uncontrollable variables is usually not very clear-cut, nor need it be. For example, even though the inventory control manager may only slightly influence sales demand, he or she may still want to consider the effect that a change in the forecasted demand will have on the optimum values of the decision variables. By this means the manager can gain insight into the vulnerability of the system to uncertain demand.

Decision models take a great variety of forms:

- A simple equation, such as the classical inventory formula

$$TVC = Cq/2 + DR/q$$

where

TVC = total annual variable cost for a given inventory item (the outcome variable)

q = order quantity (the decision variable)

C = carrying cost per unit (uncontrollable variable)

D = annual demand for the item (uncontrollable variable)

R = reordering cost per order (uncontrollable variable)

- A **mathematical programming** formulation, in which the model is represented by a formula—called the **objective function**—that computes the outcome variable as a function of the decision variables; the decision variables are confined within specified ranges by a series of **constraint** equations (e.g., total output cannot exceed plant capacity).
- A **spreadsheet** model, in which all variables are displayed in individual "cells" in a two-dimensional table, and the value of each cell can either be assigned as an input to the model (e.g., typed in from the keyboard) or computed as a function of the values in other cells (e.g., cash receipts in August = .25 × Sales in August + .70 × Sales in July + .05 × Sales in June).
- A **simulation** model, in which a computer program duplicates the logic of handling events in the real world under investigation—such as handling the flow of multiple jobs from one work location to another in a production facility; by running the model under specified decision conditions, the simulation program is then able to measure the consequences of the decisions.
- An **expert system**, in which a computer program duplicates the "intelligence" of an expert in handling a complex diagnostic or analytical task (such as setting the insurance premium rate for a fleet of trucks).

Although these models differ widely in form, all are designed to predict the outcome for a given decision. A model thus provides a means for the decision maker to explore alternative actions to search for one that meets his or her objectives. In some cases, mathematics can be employed to find the **optimum** decision—that is, the decision that results in best predicted performance according to a specified objective (e.g., maximize profit or minimize cost). In many cases, however, optimization techniques cannot be applied, and therefore the human decision maker bears the responsibility for proposing alternatives and terminating the search when an adequate alternative has been found.

A model-based DSS offers a number of advantages. An obvious one is the improvement that it may bring to a decision process. In the face of many variables, complex relationships, and uncertainty, a decision maker may have great difficulty in estimating the likely consequences of a course of action. Without such an estimate, however, there is no sound basis for a choice among alternatives. A model in such circumstances can often substantially improve a decision maker's understanding of the effects of alternative decisions and the vulnerability of the outcomes to errors in forecasting uncontrollable variables.

There are some important but less obvious advantages of using a decision model. The formulation of a model requires the organization to specify explicitly the important variables, relationships, and constraints associated with a given decision process. This can promote better communications and clearer understanding among those affected by the decisions. Decisions resulting from such explication may also be accepted as being more "rational" and "fair" than decisions made subjectively without a stated set of objectives and assumptions.

For example, a production planning model may help marketing, manufacturing, and financial executives improve their mutual understanding of the relationships among the production run size, manufacturing costs, inventory levels, and delivery performance. Similarly, a system for scheduling a university's classrooms might serve the useful purpose of defining the explicit weights given to such factors as the match between expected enrollment and the space available in a classroom, the travel distance for the students and instructor, and the rank of the instructor.

To gain this benefit, however, those affected by a decision—the **stakeholders**—must understand the substantive issues lurking in the model. A model typically embeds a number of policies, and so it is incumbent upon managers to participate in the development of the model at the policy-setting level. Within a production planning model, for example, decisions may be made regarding the allocation of production capacity among competing products and locations. Executives need to understand the model well enough to ensure that the implied policy decisions and assumptions made in the model conform to management objectives.

Successful models typically begin with an excessively simplified view of reality. As experience and understanding are gained, the model is often extended and enhanced to add new features and make it a more faithful representation of the real world. This bootstrapping process is likely to accelerate organizational learning. Unlike individual learning, however, learning incorporated in a model can stay permanently within the organization. The model thus serves as a powerful vehicle for organizational adaptation by providing an explicit, formal, and continually improving abstraction of the organization's goals, decisions, and environment.

Naturally, all of this has a cost. The development of a comprehensive model takes time, resources, and management commitment. It also takes hard work, for nothing can be more difficult than thinking deeply and explicitly about the organization's goals and decision processes.

Defining goals and assumptions in the explicit form required for a model generally serves a useful purpose, but it also runs the risk of being disruptive and divisive. For example, a decision model that uses an explicit formula to set salaries (e.g., based on some measure of contribution, experience, years with the firm, and so forth) is likely to generate a good deal of heat. A classroom scheduling program that explicitly gives a full professor preference over an assistant professor might not gain universal acceptance. Although explication of sensitive issues such as these is often valuable, and gives the organization an opportunity to decide what it wants to do and how it wants to do it, the modeler should at least be aware of the dangers involved. Sensitivity to such matters can often avoid unnecessary risks and disruptions.

Methods of Increasing Information Selectivity

We have seen that a data-oriented DSS should provide the decision maker with selected information. In designing a reporting system, one wishes to avoid two different types of error. One error is the failure to display information that the decision maker would regard as relevant and useful if it were made available. The other error is the display of useless information that is not relevant to the decision maker. The ideal system avoids both these errors, and reports all the relevant and none of the irrelevant information.

Alas, this ideal cannot be achieved. The relevance of a given item of information depends on the details of the decision process for which the information is intended. It is not possible to design a reporting system smart enough to know whether each conceivable item of information will be valuable enough to justify generating and displaying it. Users must therefore make subjective judgments about what information should be reported.

In doing this, designers have to trade off the error of displaying irrelevant information against that of failing to display relevant information. In general, reducing one of these errors increases the other. This is analogous to the errors in statistical quality control, where, for a given sampling plan, reducing the "producer's risk" (accepting bad material) increases the "supplier's risk" (rejecting good material), and vice versa.

It is possible to improve the design of a DSS so that the tradeoff curve between the two reporting errors is shifted toward a lower rate for both (analogous to using a larger or more efficient sample in a quality control system). Thus, a good DSS can simultaneously increase the likelihood of displaying useful information while also reducing the volume of useless information. This improvement can be achieved through the use of **summary reports**, ad hoc queries, exception reports, and graphical displays.

INVENTORY REPORT

23 APRIL 198X

APPLIANCE	NUMBER OF SKUs	DOLLAR BALANCE (000)	DOLLAR STANDARD (000)	DOLLAR DEVIATION	BALANCE/ STANDARD
DRYERS	37	$239	$243	($3,926)	98%
FREEZERS	17	82	102	(19,897)	80%
REFRIGERATORS	158	3,188	2,546	642,623	125%
WASHERS	42	371	371	585	100%
TOTAL	254	$3,881	$3,261	$619,386	119%

Figure 5-1. A typical summary report.

Summary Reports

The use of summary information is, of course, a widely used means of boiling down massive raw data into a more useful form. Rather than reporting individual sales transactions, for example, a sales reporting system typically provides information summarized by product grouping, geographical location, salesperson, or other meaningful dimensions of aggregation. Such a report can be displayed either in printed form or on a CRT screen.

Consider the case of the inventory report for major appliances illustrated in Figure 5-1. This gives summary information classified into each of the product categories of dryers, freezers, refrigerators, and washers. Each such category is composed of a number of "Stock-Keeping Units," or SKUs, where an SKU is an individual inventory account for a particular model stocked at a given location (a 19-cubic-foot refrigerator with a given set of features stocked at the Cleveland warehouse, say). The report shows that there are 37 SKUs within the Dryer classification, for example, which have an aggregate inventory balance of $239 thousand.

In addition to the actual inventory balance for each category, the report also gives a standard, or target, that provides a basis for judging how well inventory is being controlled. The standard for dryers, for example, is $243 thousand, which is derived by aggregating the standard inventory level for each of the individual dryer SKUs. (For our purposes here, we can assume that the standard accurately reflects the objectives of the organization, and takes account of the forecasted rate of sales, the cost of carrying inventory, the costs of reordering, and the desired level of service.)

This report may be admirably suited for some purposes, but it suffers from the fact that the aggregation process may wash out relevant information. Dryers, for example, are shown to have a total inventory balance equal to 98% of their standard ($239/$243 = .98), but this does not necessarily mean that all is well. It may simply mean that the amount of surplus inventory for some dryers closely balances an inventory shortage for others.

Figure 5-2 gives information at a similar level of aggregation as the previous report, but the categories of aggregation are defined in a way that reveals much more about the relative inventory balances. Three categories are defined: the SKUs that fall within specified control limits, those that exceed the upper limit (greater than 150% of their standard), and those that fall below the lower limit (less than 50% of their standard). The report shows, for example, that only 57% of the items are in control, while 27% of them, contributing 52% of the dollar balance, are defined as having surplus inventory. This report gives much better information than the previous one about the firm's distribution of inventory.

A summary figure may always hide relevant details, no matter how well the categories of aggregation are defined. A user may therefore wish to penetrate into the details of a summary report. Consequently, the reporting system should provide backup details for every summary figure. The reports should be "nested" in increasing detail, so that one may easily peel away the summary information to dig as deeply as necessary to find the source of a problem. For example, a user might want to see a breakdown by product category of the 68 surplus SKUs shown in Figure 5-2. If one of the categories—refrigera-

MAJOR APPLIANCE INVENTORY REPORT

25 APRIL 198X

STATUS	NUMBER OF SKUs	DOLLAR BALANCE (000)	DOLLAR STANDARD (000)	DOLLAR DEVIATION	BALANCE/ STANDARD
IN CONTROL	144 57%	$1,713 44%	$1,668 51%	$44,550	103%
SHORT <50%	42 17%	$140 4%	$560 17%	($419,951)	25%
SURPLUS >150%	68 27%	$2,028 52%	$1,034 32%	994,787	196%
TOTAL	254	$3,881	$3,261	$619,386	119%

Figure 5-2. Report summarizing by control categories.

tors, say—seems to be in trouble, the user may then want to see the details by individual SKU (i.e., by specific refrigerator model and stocking location).

A well-designed DSS can improve on the use of summary information in several ways. First, it can facilitate the definition of aggregation categories that are appropriate for a given type of decision. Using a fourth-generation reporting language, a designer can readily tailor a summary report to the needs of an individual user. (An example of this is given in the next section on ad hoc reports.)

Second, nested reports can be made available so that the user can easily determine the composition of any summary value. In Figure 5-2, for example, a user should be able to get a listing of all SKUs included in the surplus category. Although this concept certainly did not originate with DSS developers, many existing reporting systems still do not make it easy for the user to obtain backup details. An effective system should provide links upward to summary reports and downward to more detailed backup reports.

A DSS can provide interactive access to a hierarchy of these linked summary reports. The number of printed reports can grow formidably in a comprehensive reporting system, making it inconvenient for the user to locate a given one. A DSS can easily display a specified report, and make it very simple for the user to probe up and down the hierarchy of reports. For example, by merely pointing to a summary figure with a mouse, the user could call for the next layer of detail that shows the derivation of the specified value.

Ad Hoc Queries

One of users' more persistent complaints about their MIS is that the system does not generally deal effectively with unanticipated needs for information. The conventional approach to system development is to ask users to specify their information needs, which the technicians then satisfy by writing a program in a procedural language. Such a system tends to be inflexible, making it difficult to add or change reports. It is possible, to be sure, to modify the existing program when a new need is recognized, but the attendant delay and cost very often render this impractical.

Under these circumstances, the system is most unlikely to provide all the useful information potentially available. The problem stems from the fact that there exist a huge number of ways in which information could be reported, and consequently only a very small fraction of the potential reports can actually be generated on a periodic basis. In a relatively simple sales reporting system, for example, the data could be aggregated and sorted by any combination of such variables as product, customer, customer industry, customer location, sales region, sales person, profit margin, and profit contribution. It is almost a hopeless task to choose in advance, on a one-shot basis, the set of reports needed to assess sales performance and make continuing decisions on pricing, marketing strategy, staffing, and the like.

A report designed to satisfy one need is generally not very useful in meeting unanticipated needs. Take the telephone book, for example, which is a special case of a periodic report. Humans are quite adept at selecting a particular item out of a large list of items arranged in a well-defined order—alphabetical listing by subscriber name—and so manual selection in this case is quite efficient even though a telephone book has a very low density of useful information for any one user. However, finding the name of a subscriber whose telephone number is known, or all the subscribers in a given exchange, would be an entirely different matter. The standard telephone book would be essentially useless for these tasks, despite the fact that it contains all the desired information. Each new need generally requires an additional report (a listing in telephone number sequence, in this example).

If periodic reports are the only source of information from the MIS, a user might attempt to overcome the inherent limitations of this approach by specifying every kind of report for which a plausible need might arise. This obviously lowers the density of useful information generated by the system, and still provides only a tiny fraction of the potentially useful reports. Most useful management information calls for considerably more than merely listing the information in some sequence; it often involves complex selection criteria and arithmetical calculations as well. Anticipating complex specifications to meet quite specific needs is obviously not too likely. About the best the user can do, therefore, is to choose the relatively few reports that meet the most important known needs, and resign oneself to the fact that the system will probably not be able to meet many other needs.

An *ad hoc query* system can go a long way in providing a remedy for this dismal situation. The system can be designed to allow the retrieval of information that was not anticipated in detail and "wired in" to the periodic reporting system. The desired information is defined in terms of a nonprocedural query language. The query is usually entered from a display terminal, with the results made available either interactively or at a later time through a batch processing job (executed during the overnight shift, say, with the report delivered the following morning).

Interactive processing permits a trial-and-error formulation of a query, in which the results of one query can be used immediately in specifying the next one. In dealing with an unstructured problem, the user may find it helpful to browse around the database to gain a better understanding of a situation. The use of batch processing, with its greater machine economy, is appropriate for a large task that is well enough understood to formulate a one-shot query (or for which the user is willing to probe the database over a several-day period).

An ad hoc query must define for the system the source of the data for the desired report, the criteria for selecting a given record, and the specific fields of information to display. For example, in generating summary information about the inventory items within specified control limits (as shown earlier in Figure 5-2), a query could be expressed in the following way:

```
SELECT SUM(BALANCE), SUM(STANDARD)
FROM SALESDATA
WHERE BALANCE/STANDARD BETWEEN .5 AND 1.5
```

This example is a common form of query, which uses a **command language** (specifically, the language **SQL**) to define desired information. It requires that the user know such things as the name of the file from which the data are retrieved (SALESDATA), the names of the fields of information of interest (BALANCE and STANDARD), the terms to define desired tasks (SELECT), and the syntax of the language. For an experienced user, a command language of this sort generally offers an effective means of communicating with the computer. The inexperienced or infrequent user, however, may find it difficult to remember data names, commands, and language syntax.

For such a user, other forms of dialogue with the computer may be more appropriate. With a **menu** dialogue, for example, the system displays all the possible alternatives at each point in the formulation of a query (the names of all fields of data that can be retrieved from a given type of record, say); the user must then merely select among the choices presented, rather than having to know the alternatives in advance. If the number of choices is too large to display the complete menu on one screen (which is generally the case), a hierarchical set of menus can be used—in much the same way that a large restaurant menu might be broken down into a page for the appetizers, one for the main course, and so forth. This allows the user to probe through the menus to locate a specific item of interest.

For the more experienced user, the menu approach is often very cumbersome. This is especially true if the variety of choice is so large that it takes a lot of time to display alternatives and have the user choose among them. Often the best approach is a hybrid design in which a user with some familiarity with the system may selectively inhibit the display of a menu by responding in advance with the appropriate command specification.

A few query languages now incorporate a **natural language processor**, which allows an inexperienced user to formulate a query by expressing it in a conventional language, as if for interpretation by a human reader. The computer then processes the query and generates the specified outputs. If the program detects an ambiguity, or is not clever enough to infer what the user wants, it can enter into a dialogue with the user to resolve the uncertainty.

For example, a user might enter the query SHOW ME THE BALANCE AND STANDARD INVENTORY FOR ALL ITEMS IN CONTROL. If the term "in control" had previously been defined, the computer would normally have no difficulty in responding to this query. If this were a new term, the computer could ask the user to define "in control" (and then store the response for handling future queries).

Natural language processing should not be confused with **voice recognition**. In a voice recognition system, the computer is able to interpret *spoken* words. This technology is currently limited to applications requiring a fairly small vocabulary of commands and data names (a few hundred at most). The words have to be spoken carefully and discretely (i.e., with a slight pause separating them), and users are generally limited to those who have "trained" the system to recognize their particular voice intonation and inflection.

Current research in voice recognition is aimed at increasing the size of the vocabulary that can be recognized, the variety of users that can be accommodated, and the speed and continuity with which words can be spoken. A general voice recognition system requires enormous amounts of computer capacity, well beyond current economic limits. In the long term, though, this technology is likely to find increasing use throughout an MIS, including decision support systems. When combined with natural language processing, such technology provides an ideal human interface for many types of applications.

Natural language processing and voice recognition are examples of the application of **artificial intelligence** (**AI**). The great current interest and effort going into AI is likely to spawn a number of sophisticated applications in the field of decision support systems. We can expect to see a growing number of instances in which humanlike intelligence is embedded within a DSS.

It may belabor the obvious to point out that mind reading is not a likely development in artificial intelligence. Voice recognition and natural language processing will therefore not relieve users from having to define, in some form, what they want to know from the system. This is generally the hardest part of effectively employing an information system.

Exception Reporting

Exception reporting is an old principle not widely applied. The idea is to use the information system to filter out irrelevant information and present to a user only the relevant information—that is, information with a reasonable probability of leading to useful actions.

Problems with Exception Reporting A number of factors account for the relatively infrequent application of the exception reporting principle:

- Many users do not trust the "intelligence" and reliability of the MIS enough to be willing to act under the principle that "no news is good news" (in much the same way that many motorists prefer a gauge that gives a continuous reading of engine temperature, rather than relying on a warning light that identifies an "exception" when the temperature exceeds some set value).
- Most exception reporting systems use simplistic definitions of an exception—generally a proportional or absolute deviation from some standard.
- Most exception reporting systems do not make it easy for the individual user to tailor the reporting to his or her own changing needs for filtered information.

Many of these problems can be avoided by applying current information technology. The same technology used to process ad hoc queries can also function in an exception reporting system. In fact, one can view exception reporting as a scheme in which standing queries exist within the system, ready to identify conditions the user has previously defined as exceptions.

Exception reporting can be applied as broadly or as narrowly as deemed appropriate. Exceptions are almost always defined in terms of quantitative variables such as the rate of sales by item or group of items, the cost per unit of input resource or product output, or the physical units or monetary value of inventory. Virtually any variable that is measured within the system can be monitored through exception reporting.

Setting a Standard A key requirement for exception reporting is a **standard** or norm against which the system can compare actual conditions. An exception occurs if a significant deviation exists between the standard and the actual condition being monitored. If, however, the gap is less than a specified limit, the system does not report the condition. In an inventory control system, for example, the current inventory balance for an item might be compared to a desired standard level to determine if an exception exists.

Ideally, the standard should reflect the optimal value for the particular variable being measured. The inventory standard, for

Elmsford
Taxi 592.9801

Other
Taxi 948.3100

Village SB
John McKenna
914.939.7200

example, should be set at a level that achieves the optimal balance between the costs of carrying inventory, the cost of replenishing the stock, and the cost of stockouts—taking into account such matters as the current cost and forecasted usage of the item. The gap between such a standard and the actual inventory balance thus provides a measure of the penalty of being at a non-optimal level. If the deviation becomes large enough, action (presumably) must be taken to reduce the gap.

It is to trigger such action that the exception report is displayed. In principle, the exception limits should be set at the point where the expected penalty of not correcting for a deviation just begins to exceed the cost of making the correction (i.e., the management time and other costs to review the situation and make a decision, plus the cost of putting the changed plan into effect). "Good" deviations (e.g., sales exceeding the forecast) should be identified just as "bad" ones are, although the control limits need not be symmetric; too much of a good thing may cause serious problems, such as a capacity shortage if actual demand exceeds its forecast. Any significant deviation may open an opportunity to take a new action that improves performance.

In practice, the system is rarely smart enough to know the true optimum that could provide the basis for such a standard (and, if it were, the system might also be smart enough to take appropriate corrective action without human intervention). Accordingly, standards in an exception reporting system tend to be relatively static and simplistic. The following examples are typical:

- A standard is often based on an existing plan, such as an annual sales plan that includes targets by product group, or a project plan that establishes the cost and schedule for launching a new product.
- Standards might be set as a matter of policy or through the exercise of experienced subjective judgment, such as a target inventory standard of, say, three months' supply.
- A standard can be determined through an engineering approach that determines the appropriate level of resources for a task performed according to a prescribed (best) method, such

as the checkout functions at a supermarket or the teller functions at a bank.
- Historical data might be used as the standard or norm, such as the past average cost per pound for a given material, or monthly sales one year ago.
- Performance in other parts of the organization having a similar mission might be taken as the norm, such as the average cost per ton-mile for intracity truck delivery in a national distribution system.
- A standard might be computed based on a (generally simple) model—for example, an inventory standard of one month's safety stock plus one-half the **economic order quantity**, or **EOQ**.

However defined, the standard used in a particular situation should represent the best and most current definition of desired performance. Any gap between the standard and actual performance represents a source of potential improvement that may deserve management attention. As the system becomes "smarter"—as more detailed planning models are implemented, for example, or more accurate forecasting methods are developed—the new knowledge should be reflected in the definition of an exception. Adaptive improvement of standards is one of the key concepts behind an effective exception reporting system.

Exception Criteria Naturally, it is only possible for the computer to detect exceptions for those variables for which actual performance data enter the system. If, for example, the system does not collect detailed data pertaining to individual sales items at a supermarket checkout counter, then it has no basis for comparing actual performance against whatever standard might exist. As interactive terminals invade all operational aspects of the factory, office, and service facility, the computer can conceivably apply exception criteria to almost everything. With such huge masses of operational data flowing through the system, it will become all the more important to incorporate effective filters. Not the least of the issues that management must confront are the behavioral and ethical questions raised by a technology capable of monitoring everyone and everything in the organization at an unprecedented level of detail.

Tolerance limits in an exception reporting system should, in principle, be defined independently for each monitored variable. In many cases, common control limits do not provide a very effective filter for a group of variables. For example, a 20% forecast error for a low-selling item may not be at all significant, while it may be crucial for a high-volume item. In some circumstances, the fact that a variable has not yet broken through its control limits may be less significant than the fact that its rate of change has shifted significantly—such as an engine temperature gauge that suddenly starts moving rapidly toward the red danger zone.

These examples illustrate the point that more than one variable should sometimes be considered in identifying an exception. In monitoring inventory levels, for instance, exception control limits might be defined in terms of the absolute monetary deviation—over $5000, say—as well as the percentage gap. A smart exception reporting system may add greatly to the amount of data that must be monitored and the complexity of the model used to establish a standard, but in an era of relatively low-cost information processing and high-cost people, the tradeoffs increasingly favor putting significant computer intelligence into the monitoring process.

The notion of building intelligence into an exception reporting system can be extended greatly beyond conventional limits. Exploiting artificial intelligence methodologies looks quite attractive in some cases. In a casualty insurance company, for example, an AI system might assess the plausibility of claims submitted by policyholders. In doing this, it could apply a complex standard that considers such matters as the amount of the claim, the claim history of the policyholder, and any suspicious circumstances surrounding the claim (such as a rash of similar injuries reported by the same attending physician). A claim that violates this screening could be flagged for close scrutiny by a human claims investigator.

The means available for expressing exception criteria depend on the nature of the task. In most cases exceptions can be defined in a manner similar to that used in describing an ad hoc query. To define an item whose sales significantly exceed its forecasted level, for example, one could use the expression

```
SELECT ITEM_NUMBER, ITEM_NAME, SALES, FORECAST,
SALES/FORECAST
FROM SALESDATA
WHERE SALES/FORECAST>1.2
  AND SALES - FORECAST>5000
ORDER BY ITEM_NUMBER ASCENDING
```

A language of this sort gives a great deal of power for expressing criteria for filtering out irrelevant information. Since each user may have his or her own notion of what constitutes an exception, the system should allow users to tailor their reports by expressing individualized exception criteria. Furthermore, the user should be able easily to modify the criteria at will, as conditions change. Under some conditions, or at some points in time, a user may define very narrow (or zero) limits to obtain a detailed (or exhaustive) picture of what is going on; under other circumstances the user may want to apply broad limits to restrict attention to only the most egregious exceptions.

An exception reporting system should be viewed by the user as having flexible "dials" that can be set to achieve a specified degree of filtering. Often it is desirable to apply the filters interactively in trial-and-error fashion, with the filtering criteria appropriately modified according to the results from earlier responses. As users gain experience, they can become quite expert at making the system an effective tool for providing a concise picture of essential events in the real world.

In the case of embedded humanlike intelligence, such as used in the example of insurance claims processing, exception criteria are typically expressed in the form of a set of complex logical rules (e.g., if the insurance claim is for less than $1000 and no claim has been made for two years, treat as a routine loss; if the claim exceeds $50,000, review by claims adjuster; . . .). One of the languages used in AI or expert systems applications might be appropriate (e.g., **Lisp** or **Prolog**, or one of the higher-level AI languages now available in the market). The most difficult part, clearly, is defining appropriate exception criteria. The

DSS designer must generally look to experts in the field (e.g., an experienced claims adjuster), and elicit their expertise. Doing this effectively, and then representing the expert knowledge in a form the computer can deal with (called **knowledge engineering** in the AI/expert system trade) presents the most difficult challenge of applying such methodology.

As a final point, it is worth noting that exceptions can be defined in nonquantitative terms as well as in the more common form of quantitative variables or logical rules. For example, a division sales manager might wish to receive a copy of any correspondence with the Able Steel Company that contains the phrases "quality problem" or "delivery problem." As textual data become more common within an MIS—in word processing, electronic mail, "business intelligence" systems, bibliographic databases, and the like—we are apt to see widespread use of such nonquantitative exception criteria. Natural language processing could prove to be of special value in such systems. Special-purpose hardware, designed specifically for very rapid text searches, is now available and could be very attractive for applications of this sort.

Graphical Reports

The computer market now offers a variety of cost-effective hardware and software for generating graphical outputs. The technology covers a wide range of display media (e.g., CRT versus hardcopy), color variety, speed, and resolution.

Graphical information may be portrayed in a variety of ways, as shown in Figure 5-3. In addition to its general form and shape, a graph can display information through such means as color, line thickness, or **icons** that provide symbolic representations (such as a stylized picture of a truck in a vehicle routing system). In the case of a CRT display, information can also be coded by light intensity and by dynamic effects, such as flashing characters.

The typical business report consists of columns and rows of numerical information. A tabular format of this sort is quite

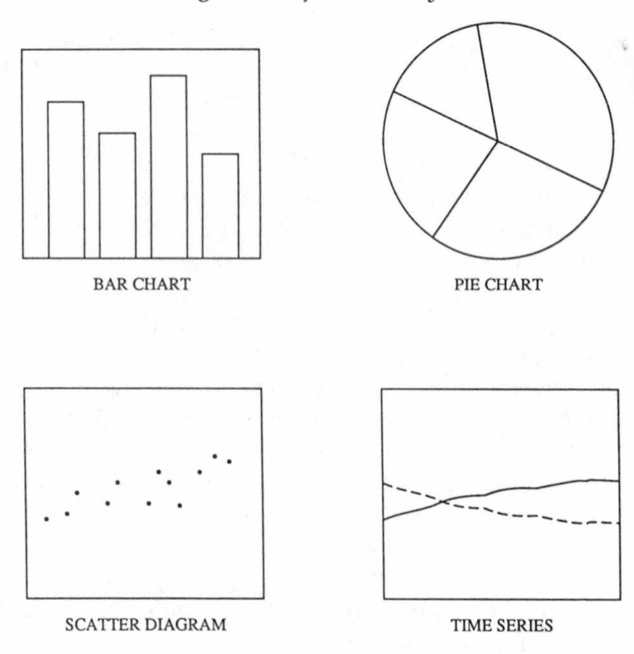

Figure 5-3. Various graphical formats.

effective in dealing with relatively simple concepts, such as the percent composition of sales by product group; this is especially the case if precise values are needed. When relationships become more complicated, however, and involve several interacting variables, a graphical display may offer the most effective way to convey information.

Consider again the case of an inventory report in which one wishes to provide the reader with a concise picture of how well inventory is being controlled for a large number of individual items. It is difficult to do this in tabular form without swamping the reader with a great deal of information that would be difficult to assimilate, or without running the serious risk of washing out useful information in the process of summarizing the data.

Figure 5-4 shows three different ways of portraying the inventory status. In Figure 5-4A, the ratio of the actual to the

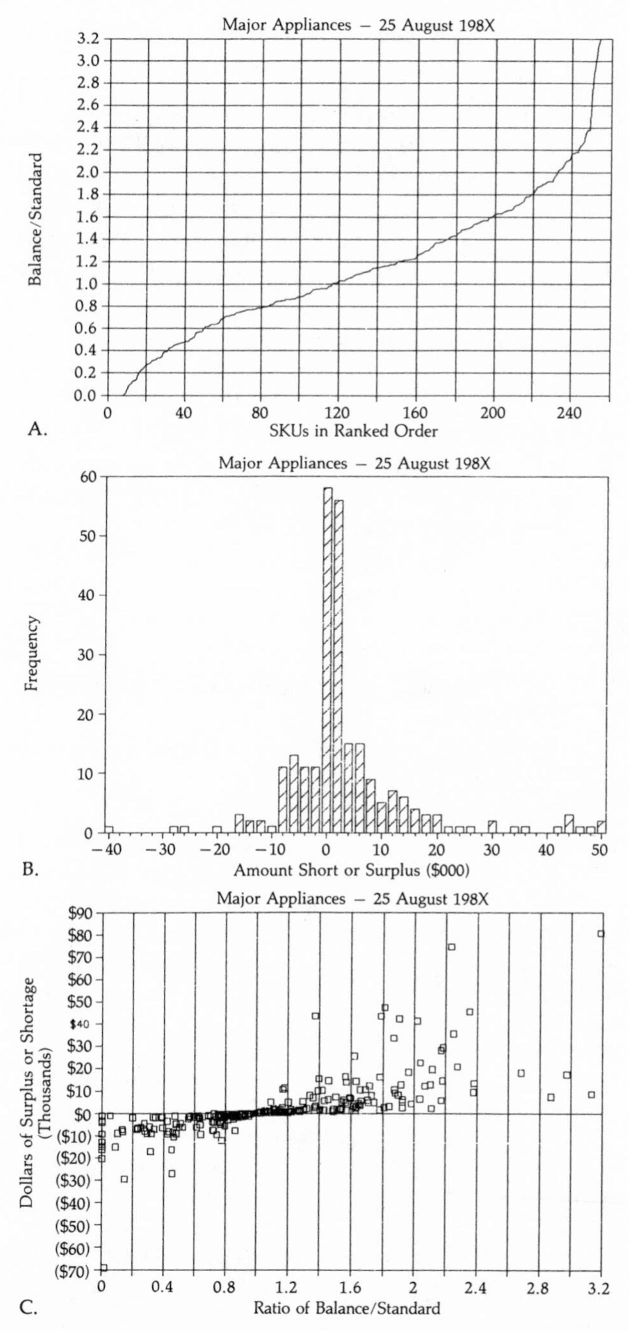

Figure 5-4. Examples of graphical inventory reports. (A) SKUs ranked by status ratio. (B) Distribution of inventory dollars. (C) Scatter diagram: ratio vs. dollars.

standard inventory balance is shown for each item in ranked order. From this single graph the reader can quite easily gain a good overall grasp of the relative stocking position of all inventory items. If attention needs to be focused on the distribution of inventory within product categories (dryers, freezers, etc.), the points on the graph could first be sorted by category, with the inventory ratios within each category then displayed in ranked order.

The inventory ratio provides a useful measure of how well inventory is being controlled, but it does not give much information about the level of investment in inventory. Figure 5-4B remedies this by showing the distribution of the dollar investment according to the amount of the short or surplus inventory (i.e., the amount by which the actual inventory falls short of or exceeds the standard, respectively).

Finally, Figure 5-4C combines information about the inventory ratio and the distribution in monetary terms. In this *scatter diagram,* a given inventory item is represented as a point in two-dimensional space, with the horizontal axis giving the ratio value and the vertical axis showing the investment level relative to the standard.

These examples illustrate some important general points about graphical displays:

- There is no one "right" way to display information: a variety of methods can be used to reveal different aspects of the same basic data, or to highlight points of interest to different users.
- The design of a graphical display is much more an art form than a science: it takes a creative act to think of an effective way to boil down masses of interrelated data and convey their essence in a useful and concise form.
- Under the right circumstances, a graphical display can provide a much more effective way of portraying a situation than can equivalent tabular information.
- Individuals may have quite different attitudes about how they want information displayed, and so an effective system must cater to the idiosyncratic needs of each user.

Graphical display has often been trivialized, with fancy charts used when a simple tabular report would do at least as well (as in the case of using a multicolored pie chart to show the composition of a simple aggregation). Furthermore, without suitable care, a graphic display can sometimes distort rather than elucidate. At their best, though, graphs can be extremely illuminating—for example, when multiple interacting variables are portrayed in a way that makes intuitive sense to the recipient. Color and other forms of coding can be useful when they add additional dimensions to the information being represented, rather than merely tarting it up.

Despite the great intuitive appeal of graphical display, surprisingly little hard evidence exists to support its use. Some experiments seem to confirm the value of graphical output, while other research points in the opposite direction (or comes to ambiguous conclusions). At the very least, though, we can state with some confidence that graphical output is not a universal solution to effective reporting. Users' attitudes toward graphical display are highly individualistic: some like it a great deal and make excellent use of it, while others cannot tolerate it. Even when a graphical display proves of great value, it may need to be supplemented with "real" data giving precise numeric values for critical points on the graph. The DSS should make it easy for the user to obtain such data—for example, by pointing on a CRT screen with a mouse.

Optimizing Models

The use of models has roots going back to World War II (and even before, if one counts some of the early work in industrial engineering at the turn of the century or the inventory models developed in the 1920s). The field of **operations research**, which largely overlaps the field of **management science**, deals with the application of mathematical models to decision making.

Operations research specialists traditionally aim at the development of an **optimizing model** to solve a given problem. Such a model has as its purpose the identification of the best possible solution for the problem in question, without involving a human decision maker directly in the optimization process (once the model has been formulated). In a production scheduling problem, for example, a model might be used to find the schedule that minimizes production costs or maximizes total profit contribution within capacity constraints.

There are several reasons why a discussion of decision support systems should consider the role of optimizing models:

- It is useful to consider an optimizing model as a special case of a DSS—one in which the model provides a particularly high degree of aid for the human decision maker.
- A non-optimizing DSS may include lower-level optimizing models that deal with subproblems, such as a budgeting DSS that uses a model to optimize production schedules subject to various budget constraints.
- Any optimizing model in fact provides only an approximate solution; a model in the hands of a human decision maker exploring the effects of the approximations (called **sensitivity analysis**) thus constitutes a DSS.
- Involving a human decision maker as a direct participant in an optimization process can sometimes lead to a better solution or a more efficient optimizing procedure.
- An understanding of the strengths and weaknesses of optimizing models provides some useful insights into the more general role of decision support systems.

Requirements for Optimization

Optimizing models can only deal with well-structured problems that meet three quite restrictive conditions:

1. The problem must be understood well enough to permit a quantitative measurement of all important variables, along with an explicit quantitative definition of all important relationships and constraints among these variables.
2. An explicit, single goal—called the **objective function**—must be defined so that its value can be computed for any alternative solution.

3. A computationally feasible procedure (algorithm) must exist for finding the optimum value of the objective function.

These requirements restrict optimization to a relatively narrow range of problems. Consider, for example, the difficulties one would face in developing the optimum annual operating budget for a manufacturing firm. Any serious attempt at optimization for the firm as a whole would require an explicit knowledge of the quantitative effects of expenditures on such wide-ranging activities as research, advertising, and safety programs. It would be patently absurd to expect stakeholders within the firm to reach any sort of consensus about these relationships.

It is not enough to understand how decisions relate to outcomes; optimization also requires that all outcomes be boiled down to a single objective function. (Contrary to frequent advertising claims, it is not possible to provide the highest quality product *and* the lowest possible cost.) In establishing an annual operating budget, management must consider such factors as earnings per share, stability of earnings, market share, long-term programs aimed at product improvements or lower costs, and employee safety. These goals generally conflict, and it is rarely possible to establish a formal tradeoff among the goals (between profit and safety, say) to derive a single composite objective function.

Finally, even if it were possible to develop a satisfactory model and an objective function for the budget—a pretty heroic assumption!—the resulting model would probably not be susceptible to optimization because of its complexity. Even with the huge increase in the power of computers, and improved mathematical techniques that significantly increase the efficiency of optimizing algorithms, there still exists a wide range of models for which optimum solutions cannot be computed. For example, despite the relative simplicity of the game of chess (compared to budgeting a large enterprise), finding the *optimum* chess move is computationally infeasible (although the computer can play an impressively *competent* game when programmed using artificial intelligence techniques).

The requirements for optimization are obviously difficult to satistfy, and so relatively few decision problems can be optimized. In fact, it is *never* possible to find the true **global** optimum—that is, the optimum set of values for all possible decision variables over which the organization has some degree of control. Strictly speaking, the best that can be done is to find a **suboptimal** solution that determines the values of only a few decision variables while treating all others as fixed constraints. By thus severely limiting the boundaries of a problem, a model can be simplified enough to make it mathematically tractable. An inventory model, for example, might determine "optimum" order points and order quantities, holding constant the aggregate production capacity and detailed scheduling rules (which, for a truly global optimum, would have to be treated as decision variables).

In addition to restricting their scope, models are further reduced in complexity by using aggregate variables in place of their more detailed components, substituting simple functional relationships to represent more complex ones, and treating all variables as **deterministic** instead of **probabilistic**. A particularly common simplification is the use of **linear** (straight-line) relationships in which the effect of a change in a variable is strictly proportional to the amount of the change.

Complexity not only increases the difficulties of model development and data collection, but also adds to the cost of computing the optimum. Linear models enjoy great popularity because very efficient algorithms exist for finding the optimum of a linear model. Optimization may very well not be feasible without the use of linear approximations.

Heuristics can often be employed to reduce the cost of computing. A heuristic is a simplifying "rule of thumb" built into an optimizing algorithm to reduce the number of alternatives that must be considered. For example, an algorithm for finding the optimum routes for a fleet of delivery trucks might arbitrarily combine the deliveries for customers within one mile of each other, even though it can happen that such customers might sometimes be more economically serviced from different routes. The heuristic would be justified if the resulting routes closely approximate the true optimum routing (if indeed the real optimum can be found). The inclusion of such heuristics can easily spell the difference between a computationally feasible algorithm and an elegant but impractical one.

Model-Oriented Decision Support Systems

Characteristics of a DSS Model

Unlike a conventional optimizing model, a decision support system depends on human judgment as an integral part of the decision process. A computer-based model can provide invaluable aid in predicting the consequences of a given alternative, but the creative act of generating alternatives, and choosing among them, generally remains the responsibility of the human decision maker. By relying on a human to perform the unstructured tasks that cannot be included in a formal optimizing algorithm, DSS methodologies substantially expand the class of problems susceptible to modelling. A DSS model can readily deal with nonlinear probabilistic functions, for example, which are generally beyond the capabilities of optimization techniques.

A DSS model is not restricted to problems having a single objective function. The decision maker usually confronts multiple decision criteria, and looks to a DSS model to predict the consequences of a proposed action in terms of the various criteria of interest. In evaluating a proposed annual budget, for example, a DSS model might predict such matters as earnings per share, market share and profit margins by product group, employment levels over the year, and borrowing requirements.

Use of a Model-Oriented DSS

When armed with a DSS model, decision makers seek to generate an acceptable plan through an exploratory search process. A decision maker typically enters a proposed plan, defined in terms of such decision variables as production levels, resource allocations, and financing sources. The computer then uses the model to predict the consequences of the plan. Upon reviewing the predicted outcomes, the decision maker generally revises the plan in some way to improve the expected results.

Through a trial-and-error process of this sort, alternatives can be evaluated until one is found that is deemed to be acceptable or until the search is abandoned. This process is not at all likely to discover the optimum plan (even if one could define it), but that, of course, is not the issue; the real issue is whether the DSS improves the decision process enough to justify the cost of developing and using the system. Often it can.

A model typically lies at the heart of a sophisticated DSS, but the system might also include the following functions:

- Trigger the decision process through an exception reporting system that identifies possible problem areas, or through a "tickler" file that notifies the appropriate decision maker when a given periodic decision becomes due.
- Supply selective retrieval of data for use in a model.
- Provide an effective human interface—menus, graphics, etc.—to enable a decision maker to interact effectively with the DSS.
- Aid the user in choosing the best alternative, through such means as a graphical display that shows tradeoffs among alternatives (between inventory cost and service level, say), sensitivity analyses over a specified range of an input variable, and user-assigned weights for different outcome variables.

Approximations in Models

As we have seen in connection with the discussion of optimization, a model is merely an abstraction of reality. Its value as a decision aid comes from its ability to represent behavior in the real world. A tradeoff exists between the realism of a model and the cost of developing it and providing it with data. The real art of successful modeling is the ability to capture the essence of a situation without going into irrelevant detail.

The faithfulness with which the model represents reality depends on a number of factors:

1. The accuracy of the forecasted values of uncontrollable variables (e.g., sales demand, material costs, interest rates, production times).
2. The quality of the data used to define functional relationships (such as the change in demand caused by a change in price or advertising expenditure).
3. The "granularity" of the variables used in the model—whether they represent large aggregates of things (e.g., total production in a month) or fine details (the production of an individual product on a given day).
4. The complexity of the relationships among variables included in the model—e.g., simple linear relationships, or complex nonlinear functions (such as an abrupt change in cost when a second shift is added to the factory schedule).
5. The treatment of **uncertainty**—for example, whether a single deterministic estimate is used for a variable with an uncertain value (e.g., the prime interest rate twelve months in the future), or whether the uncertainty is represented by a probability distribution (e.g., one that explicitly recognizes the relative likelihood of different interest rates).

Increasing the realism of a model generally permits the user to make better decisions. It may, however, add significantly to the cost of developing and using the model. The real question is, of course, how much does it cost to add greater realism, and what payoff does the greater realism yield in improved outcomes? The inevitable economic law of declining marginal returns eventually sets in: after some point, increasing the realism of a model increases the cost of the model more than it improves performance.

An important component of this tradeoff is the cost of collecting and maintaining data to support the model. Costs can grow rapidly with demands for finer granularity, greater timeliness, and increased accuracy of the data, especially if large volumes of data must be collected specifically for use in the model. Even in the case where the data are derived from a transaction-fed database, significant costs can be incurred in selecting and processing the data to transform them into the form needed by the model.

Evolutionary Development of a DSS Model

A successful DSS almost always evolves through a long learning process. At the beginning of the process, a "quick-and-dirty" model might be developed using a high-level language such as Lotus 1-2-3. As users gain experience and confidence with the model, they inevitably want to expand the range of issues included in the model, gain access to remote databases, generate new types of reports, and improve the realism of the model. These evolutionary changes are such a natural part of the learning process—truly one of the main advantages of using a DSS—that the absence of change strongly suggests that the model is probably not being used effectively.

The evolutionary development of a model can be aided substantially by sensitivity analyses that measure how changes in controllable and uncontrollable input variables affect the predicted outcome variables. If an outcome is shown to be sensitive to a particular input variable, the modeler can concentrate on improving the realism of that portion of the model. Sensitivity analyses not only provide information useful in enhancing a model, but they also give managers invaluable insights about the world being modeled—about what is important, the robustness of a plan with respect to changes in prices and markets, how outcomes respond to changes in decisions, and the like. The added understanding that a manager gains by such analyses can by itself justify the effort of using a model.

To adapt a DSS model as the organization learns, a model builder needs to employ development tools that facilitate change. Most models are programmed in a high-level language specialized for DSS applications. These languages typically include many of the following features:

Defining mathematical and logical relationships among decision and outcome variables.

Statistical calculations, such as mean, standard deviation, and regression analysis.

Common business functions, such as calculating depreciation charges and net present values.

Report formatting, including graphical outputs.

Calendar calculations, such as determining the number of days between two specified dates.

Consolidation of two or more submodels (e.g., division models consolidated into an aggregate corporate model).

There now exists a wide variety of software products aimed at the DSS market. These vary considerably in power, ease of learning, ease of use, hardware requirements, and cost. At the low end of the spectrum—in cost if not necessarily in functionality—are the spreadsheet languages run on inexpensive microcomputers. High-end software products designed for mainframe computers are used in dealing with very large problems requiring massive databases and sophisticated functions. They may sell for well over $100,000. An organization should almost always begin with a low-end system to acquire relatively inexpensive experience. A more sophisticated system can later be developed using more expensive tools if the expected additional benefits justify the additional cost and complexity.

DSS and Office Information Systems

As we have already discussed, many leading organizations are moving deliberately toward a distributed environment in which a high proportion of employees at all levels have ready access to a personal workstation. In addition to stand-alone DSS applications, we can expect to see growing use of workstations for accessing remote databases, software, and specialized hardware resources. The DSS will become increasingly integrated with an **office information system** that provides many of the following services:

- Text processing with integrated graphics.
- Document preparation, including high-quality printing.
- Document distribution.
- Electronic mail.
- Time management (e.g., schedule meetings).
- A variety of "desktop" functions (e.g., a corporate telephone directory with integrated dialing, electronic memo filing, and calculator functions).

Ideally, "seamless" links will exist among these various functions so that they all appear to the user as an integrated set of services with a uniform and consistent human interface. In such a system, no clear-cut boundary exists between a decision support system and the rest of the MIS.

The productivity and work quality of managers and professional workers are becoming critical issues for organizations. The hallmark of a sophisticated organization is a heavy reliance on the indirect functions performed by its white-collar staff. As a result, the success of such an organization depends increasingly on the effectiveness and efficiency of these "overhead" personnel.

An information system that provides powerful DSS and office services to support the professional staff can significantly enhance its productivity. Professional workers deal almost exclusively with information in all its forms—communications, planning, decision making, and coordination. Information technology offers the potential to make major improvements in these activities.

Effective communication is certainly a key element in such improvements. Interpersonal communications can take the form of text, images of all kinds (pictures, graphics, legal documents, signatures, etc.), and voice messages. It is now technically and economically feasible to provide virtually instantaneous communications throughout the organization, even on a worldwide basis. An advanced communication system can keep track of the identification and location of everyone formally connected to the network, making it trivially easy to reach anyone on the network (e.g., by simply addressing a message to WIGGLESWORTH @ LONDON). It does not stretch the imagination too much to envision a "smart" network that maintains a profile of the expertise and interests of each participant—allowing, for example, the broadcast of a query to everyone having experience with a specialized product or industry.

The capabilities provided by an advanced office information system seem quite likely to bring about profound changes in the management of organizations. A system that provides rapid interpersonal communications reduces reliance on structured hierarchical links. Organizations that heavily use **electronic mail** have already observed a shift to relatively informal direct links between those needing information and those who have it, often jumping over several formal reporting levels. An organization really able to marshal the knowledge and experience of its employees through a sophisticated communication system stands to gain a tremendous competitive advantage. Unfortunately, we are a long way from fully understanding the implications of our new-found technical capabilities, or how we might use them most effectively.

Further Readings

Alter, Steven L., *Decision Support Systems: Current Practice and Continuing Challenges,* Addison-Wesley, 1980. A comprehensive introductory discussion of DSS.

Anthony, Robert A., *Planning and Control Systems: A Framework for Analysis,* Harvard University, 1965. A classic book on organizational planning and control systems.

Bass, Bernard M., *Organizational Decision Making,* Richard D. Irwin, 1983. A good conceptual discussion of the important behavioral issues in organizational design and decision making.

Emery, James C., *Organizational Planning and Control Systems: Theory and Technology,* Macmillan, 1969. A discussion of planning and control from a systems point of view.

Guth, William D. (ed.), *Handbook of Business Strategy,* Warren, Gorham

& Lamont, 1985. A collection of articles by a number of leading experts, with good discussion of strategic planning and control systems.

Mason, Richard O. and E. Burton Swanson, *Measurement for Management Decision*, Addison-Wesley, 1981. One of the few publications that gives sufficient attention to the critical topic of management measurement and reporting systems and their behavioral effects.

Rockart, John F. and Christine V. Bullen (eds.), *The Rise of Managerial Computing*, Dow Jones-Irwin, 1986. A collection of a number of significant papers on decision support systems.

Sprague, Ralph H. Jr. and Eric D. Carlson, *Building Effective Decision Support Systems*, Prentice-Hall, 1982. A sound discussion of the design and implementation of DSS.

Simon, Herbert A., *The New Science of Management*, Harper & Row, 1960. A classic book by one of the leading thinkers about organizational planning and control systems.

Tufte, Edward R., *The Visual Display of Quantitative Information*, Graphics Press, Cheshire, CN, 1983. An interesting and enlightening discussion of how to improve the effectiveness of graphical presentation of information.

Winston, Patrick Henry, *Artificial Intelligence*, Addison-Wesley, 1984. A good, but demanding, textbook on artificial intelligence and expert systems.

❖ 6 ❖

Development of Application Software

Management Issues in Application Software Development

Problems with Application Software

The technical problems that an organization encounters with its MIS generally stem from difficulties in developing application software. Symptoms of the difficulties are everywhere:

- Applications too often suffer from missing functional capabilities, poor human engineering, unnecessary complexity, inflexibility, and unreliability.
- Most organizations have a two or three year backlog of software development projects that have been identified as worthy of implementation; the "hidden backlog" of projects not yet explicitly identified is undoubtedly even longer (probably infinite, owing to the insatiable demand for information services).
- Software development projects are notorious for being delivered late, at excessive costs, and with missing capabilities.
- The organization sticks with obsolete programs—often over ten years old and designed to meet the needs of a different era—because of the perceived high cost and disruption of developing new programs.
- Programmers and analysts devote a high proportion of their

time—often 75% or more—to maintaining old software; the resulting software quagmire leaves few resources for the development of new applications that properly exploit current technology.
- Competent programmers and systems analysts are difficult to find and expensive to support.
- The user community increasingly tries to escape from the problems associated with the central development staff by developing systems themselves on personal computers and on departmental minicomputers—not always in the best interests of the organization as a whole.

The Classical Approach to Software Development

It is not difficult to understand the source of software problems; they are largely intrinsic to the programming technology and organizational approach used in conventional software implementations. Software development **projects** vary tremendously in their size, technical characteristics, and management approach, but a core application usually has most of the following characteristics:

1. It runs on a mainframe computer.
2. It is designed and implemented by professional systems analysts and programmers.
3. It is written in a third-generation procedural language—generally COBOL for a business application and Fortran for a scientific or engineering application.
4. It consist of several thousands of lines of program code, and perhaps even hundreds of thousands.
5. A seemingly slight error in any one of the lines of code can cripple the program or cause serious damage due to erroneous outputs from the program.
6. One portion of the program may interact with various other parts, so that changes or errors can have widespread and unforeseen ramifications.
7. The productivity of programmers who develop the application

is typically limited to a few dozen lines of fully tested code per day.

8. Any changes in the application software to accommodate new needs must take place without disrupting the normal routine of the organization—something akin to installing a new rudder on a 747 in flight from London to New York.

Success in developing a large application program calls for a high order of technical and organizational skill. Management of these software projects requires close coordination of the activities of perhaps dozens of persons working on many interrelated tasks extending over a period of months or even years. The field of **software engineering** is concerned with the theory and practice of software development, with the objective of making the process more efficient and the resulting programs more reliable, flexible, and maintainable.

Life-Cycle Development Process

A key concept in software engineering is the staged **life-cycle development process**. A large project is far too complex to be managed in a monolithic way. Consequently, the project is broken down into a number of more-or-less distinct stages. Each stage can then be managed as a separate task, with a well-defined schedule, task assignments, resource requirements, and end products.

A variety of terms have been given to the various stages in the life-cycle development process, and the boundaries between stages may vary somewhat on different projects or within different organizations. The following breakdown captures the essential idea, however.

Feasibility Study

A **feasibility study** involves relatively broad and quick screening of a potential application to obtain an approximate estimate of its costs and benefits. An application having attractive cost-benefit characteristics can proceed to the next development stage; those that do not can be cancelled before the organization sinks a major investment into the effort.

A feasibility study should generally take no more than a few months nor consume more than perhaps 10% of the full implementation cost of an application; often it requires considerably less in both time and money. The study should be sufficiently thorough, however, that the resulting estimates of costs and benefits provide a sound basis for judging whether or not to continue the project. It is not unreasonable to expect, for example, that these early estimates err by no more than 25% of the actual values if the project were to proceed to full implementation. To get acceptable accuracy within the limited time available, a feasibility study should be staffed with a small team having a high level of skill and experience.

It is essential that the team also have a sound understanding of the organization's business needs, because an application's feasibility depends at least as much on its organizational and operational attractiveness as on the ability of the technicians to put the system together at an acceptable cost. A steering committee, composed primarily of representatives from user groups, can be very helpful in assisting the team and assessing its work.

Needs Analysis

If a proposed application survives the initial screening process, it is then subjected to a more detailed **needs analysis** to determine its desirable functions and outputs, along with such technical specifications as response time, flexibility, accuracy, security, and reliability.

Requirements analysis is often used to describe this stage of the development process. The term is misleading, however, because it implies that a user *requires* a given set of characteristics to carry on the functions of the organization. In almost all cases, though, a great deal of latitude exists in specifying the characteristics of a system. Even in the case of legally required outputs (tax or regulatory reports, say), considerable freedom may exist regarding such matters as the timing, content, and format of a report. In most cases, a "requirement" is entirely discretionary, and depends only on the tradeoff between the value and cost of satisfying it. It would never make economic sense for the technicians to go off and obediently build a system that meets every user's wish.

In this early stage of development, the cost of meeting a specification cannot be estimated with great accuracy. Furthermore, users—who presumably should play the dominant role in setting functional specifications—generally harbor only a vague notion of how much it costs to satisfy a specification; their forte is to deal with the *value* side of the equation. Users should therefore provide a *range* of needs, along with some indication of the relative value of meeting them.

One method of eliciting such information is to invite users to specify each of their needs in terms of three categories:

1. "Must do": a rock-bottom need required to carry out the essential functions of the business (such as processing customer orders within three days, say).
2. "Should do": a need that the user judges both desirable and feasible, but is willing to negotiate in striking the best balance between value and cost (processing customer orders overnight, for example).
3. "Nice to do": a "frill" that the user regards as desirable from a business standpoint, but not necessarily justifiable in terms of the cost of attaining it (on-line order processing, with immediate response to customers entering orders by telephone or their own terminal).

In the examples cited above, each level of need deals with the same basic issue (the processing time of customer orders), but with increasingly demanding specifications. In other cases, the more demanding needs may introduce entirely new functional capabilities. For example, in an order entry system, a "nice to do" request might be a new customer database that permits the analysis of past purchases to provide a more focused approach to direct mail advertising.

System Design

The purpose of the **system design** stage is to convert a set of specifications into a technical description of a physical system that meets the specifications. An important task in this process is choosing the structure of the system—the definition of its various parts and how they link with one another. As already noted, a well-designed system has a hierarchical structure in which each high-level part is broken down into finer subparts. As the design process proceeds, each component gets broken down into finer and finer detail. Eventually the point is reached where the specifications can be turned over to programmers for implementation in a computer language.

During the design process, analysts must consider **tradeoffs** among the various specifications. Improvements in the quality of the system, such as reducing the time to process customer orders, generally increases both its value and cost (although with good design it may be possible to find a more *efficient* system that improves value without increasing cost). Designers try to home in on the combination of specifications that yields the best overall balance between value and cost.

A satisfactory design generally satisfies all the organization's "must do" needs (although even these may be dropped or modified in light of further analysis). Most "should do" needs can usually be satisfied, and even some of the "nice to do" requests can be justified. With the continuing advances in information technology, users often have a poor calibration of the tradeoffs between value and cost, generally erring on the side of excessive conservatism. A user may judge an interactive system to be an unjustifiable luxury, for example, whereas today's technology may bring the cost down to the point that such a system becomes quite feasible.

The product of the design process is a detailed set of technical documents. The purpose of these documents is to describe the system in sufficient detail to permit programmers to implement a set of programs that satisfies the selected specifications. The documents define the components and hierarchical structure of the system, data inputs and their sources, the outputs, and the algorithms for transforming inputs into outputs. The structure of the database—its components and the links among them—is also a critical part of the design documentation.

Opinions differ on the degree of detail that should be included in the design documentation. Some specifications leave few options to the programmers, while others leave them considerable discretion in dealing with unresolved details. In any case, however, the design should be detailed enough to reduce to an acceptably low level the risk that programmers will encounter unpleasant surprises. A complex algorithm critical to the system, for example, should be worked out during the design stage.

Programming

Programming creates an executable computer program that achieves the functions and characteristics established in the design documents. A programmer expresses the specified procedures in the form of a computer language such as COBOL. The resulting **source program** can then be translated automatically by the computer into an **object program** that will perform the specified functions and meet the specified characteristics of the design.

Programming adheres to the hierarchical structure of the design. A large program consists of many subprograms, which in turn are broken down into still finer parts. This process continues until each of the lowest-level subprograms performs a relatively narrow and isolated function. The program resulting from this hierarchical factoring has a **modular structure**, with each component termed a **module**.

It is generally considered good programming practice to keep each module to a modest size—no more than about 60 lines of code, say, the number that can fit on one printed page. The small size of a module makes it relatively easy to understand, debug, and change. Furthermore, the relative isolation of each module restricts the effects that a change in one part of the program will have on the other parts. Without careful discipline of this sort, a seemingly minor change could have unforeseen ramifications throughout the entire program.

Given an adequate design, programming is generally a fairly straightforward process. If the design documentation is very detailed, programming—or **coding**, as it is often called—is almost a mechanical process of translating from one form of expression (the design documents) into another (the computer language). Even if the specifications leave some discretion to the programmers, the programming stage should not cost more than perhaps 15 to 20% of the total implementation cost; use of a powerful 4GL could cut this figure to well under 10%. In fact, some high-level, nonprocedural languages effectively eliminate conventional coding by providing an efficient means of automatically translating a design specification expressed in the 4GL.

A critical accompanying work product from the programming stage is the detailed **documentation** of the program. The program listing—the printed version of the source program—should embed generous "comment" statements that describe to the human reader the logic and assumptions of the program (but do not affect the translation of the program itself). Other documents are typically used to amplify the program logic and to describe such matters as the program's overall structure, the database structure, inputs and outputs, variable names, and error-handling procedures. Preparing good documentation is expensive, time consuming, and boring to most programmers, but without it program testing and maintenance are exceedingly difficult, expensive, and error prone. It's a question of paying up front for decent documentation, or paying the continuing costs of trying to maintain inadequately understood code.

Testing

A product as complex as a large computer program requires very careful testing to see that it performs as designed. Such testing may consume a major share of the resources required to implement a program—as much as half of the total cost in some cases, and almost always more than inexperienced intuition might suggest.

Despite disciplined efforts to eliminate errors, undetected **bugs** almost always lurk in a newly deployed program—and may persist even after the program has been in operation for several years. Designers and programmers typically spend a great deal of effort to anticipate all possible inputs to a program, and then test to see that the program works under the expected conditions. But testing cannot try all combinations of conditions under which a program might be run. It is always possible—indeed likely—that some (rare) combination of events can cause a previously unrevealed bug to turn up during a program's operational stage. Mature programs that have been continuously improved and widely used are usually quite reliable, but even here a bug can suddenly appear (often because it was introduced in the process of debugging or "improving" the old program).

Testing takes place as the programming proceeds. Each module is first tested by itself in a so-called **unit test**. When given a set of inputs, the module is tested to determine if it generates the expected outputs. If it does not, the source of the difficulty is identified and corrected.

After undergoing separate unit testing, a group of related modules is then assembled into a larger aggregation for testing. Such a test often reveals errors arising from interactions among the modules. Any discovered error must, of course, be corrected before continuing further. This generally entails making modifications in one or more constituent modules. The deeper one must penetrate into the lower-level modules to fix a bug, the more expensive and time consuming the correction. An important axiom of testing, therefore, is that one should focus on the detection and correction of errors at as early a stage as possible.

The testing process continues, with increasingly larger aggregations of components, until eventually the full system is tested as a whole. System-wide tests should subject the system to a realistic set of operating conditions to increase the probability that hidden bugs will show up. Certain errors occur only when the system encounters stresses due to a high rate of transactions, and so an effort must be made to simulate peak-load conditions. This is particularly difficult to do in the case of an interactive system having transactions that arrive randomly from many dispersed users.

Since the client for whom a system is implemented must eventually take over responsibility for its operation, he has the strongest motivation to see that the system works. The prospect of inheriting a faulty system in a fortnight concentrates the mind wonderfully. The client should therefore play an early role in the testing process, making sure that the conditions under which the system is tested are as realistic and thorough as possible. Some of the operational personnel who will actually use the system should also be involved.

Deployment

Following its careful testing, the system is made operational through a series of **deployment** steps. A critical part of this process is creating the database to support the new application. In some cases this may simply call for converting existing files into the database format required by the new program (which often requires, unfortunately, substantial "clean-up" of errors in the old data that were not detected in the previous applications). In most cases, however, at least some entirely new data have to be collected, entered into the system, and edited for accuracy.

If a new application has no links with existing applications, deployment is generally quite straightforward. Much more often, though, a new program has substantial interfaces with other programs. A new production scheduling application, for example, would typically have to exchange data with such applications as order entry, inventory control, distribution, and accounting. Establishing such links is often one of the more difficult aspects of implementation, possibly requiring major modifications in existing programs.

Deployment also requires attention to such matters as preparing user manuals, training the staff to use the system, and providing users with the necessary terminals and other physical facilities. Scheduling and coordinating all the interrelated tasks associated with the introduction of a large system presents a formidable management challenge. Many organizations have found it essential to employ for this purpose a disciplined project management methodology, such as **CPM** or **PERT** (and, indeed, in managing the other stages of the implementation as well).

Operation

Eventually the system becomes operational. The step of handing over responsibility to operational management should be recognized explicitly through a formal **sign-off** procedure. By agreeing to the operational status of the system, operating managers accept responsibility for making it work. The inevitability of this transfer of responsibility provides a powerful incentive for the managers to play an active role throughout the implementation process.

A system should almost always become operational through a gradual and cautious conversion process. Management should avoid like a plague a sink-or-swim situation in which the organization suddenly finds itself heavily dependent on the proper functioning of a new system. Rather, the system should be introduced in such a way that a major failure will not cause an unavoidable disruption. Although competent testing can substantially reduce the risk, it cannot guarantee that such failure will not occur: some problems only crop up in a real operational environment. Backup procedures must therefore be in place to take over if serious problems arise.

If the new system replaces an old one, the two can sometimes be operated in **parallel** so that the old system backs up the new. The initial operation of the new system can be viewed as an extension of system testing, except that live transactions replace simulated events. If a major failure occurs, the organization then reverts to the old system. Generally it should be necessary to operate both systems for only a short time before managers can feel reasonably confident that the potentially catastrophic surprises are behind them.

Parallel operation has its limitations, however. For one thing, it can be quite expensive to operate two systems, especially if a large operational staff must be dedicated to each. More fundamentally, a new system often provides completely new functions for which no existing system can serve as backup. The actions taken by the two systems—their handling of a stockout situation in an order entry system, for example—may not be compatible in a way that allows a simple reversion to the backup system if a problem occurs.

Other deployment approaches can be used in addition to, or in place of, parallel operation. For example, if the system is designed to serve multiple geographical locations or groups of users, it should first be brought up at one site only. Sufficient time should be left in the implementation schedule to allow for any corrections and modifications indicated by the experience at the first site. Ideally, all subsequent deployment at multiple sites should continue in serial fashion, with necessary modifications made before proceeding to the next site.

This same approach can be used in the serial deployment of functional capabilities. A core capability can first be implemented, sufficient to serve only the minimal operational needs of the application. Extensions and "bells and whistles" can subsequently be appended to the core after the initial implementation problems have been solved.

A serial approach to implementation takes maximum advantage of the learning that comes with real operational experience, for which even the best design and testing methodologies cannot substitute. This step-by-step process may stretch out the implementation, but it can result in a much more successful system. It might even save time in the long run by avoiding the need to make subsequent major revisions in hastily deployed and flawed programs.

Maintenance and Adaptation

A system is never really finished; it goes through continual changes over its entire life. These changes are necessary to correct bugs, adapt to new technology, and respond to new organizational needs. Program maintenance deals with these changes over the operational life of the system. The maintenance function can easily consume up to three-quarters of the total life-cycle development cost of a typical application program.

Many maintenance changes involve only minor modifications to a program. A user may request a new report, for example, or a cosmetic change in an existing report. Generally such a change does not require a modification in the program's input data or processing logic, but only takes existing data and rearranges them in a different way. Nevertheless, even a minor change in a conventional program may require help from the professional technical staff, with attendant delays and red tape. (**End-user programming**, discussed in the next chapter, seeks to reduce users' dependency on the technical staff for making these minor changes.)

Changes can be of a more substantive nature, of course. As users gain experience with a system, they soon see a number of things they would like to change—not necessarily because of a failure to correctly anticipate their needs, but more often as an integral part of the organizational learning associated with the application of information technology. It is not an exaggeration to state that *no* program ever works entirely satisfactorily until it has gone through a period of adaptation, and no organization can ever understand a system until it has lived with it through a considerable learning process. Change in a conventional system is difficult and expensive, but there is no way to avoid it.

Improvements in the Conventional Development Process

Strategies for Improving Software Development

The complexities and problems of conventional software development have important management implications. Unless substantial improvements can be made in the development of application software, organizations will find it increasingly difficult to implement responsive, cost-effective, and strategic information systems. Fortunately, a number of different approaches are available that are already beginning to have a major impact on the development process:

- Improvements in the conventional development process.
- Use of purchased application packages.
- Use of fourth generation development tools.
- Prototyping.
- End-user programming.

These various approaches are by no means mutually exclusive. Most organizations will continue to develop or purchase some core transaction processing systems based on conventional languages and implementation methodologies. The bulk of the new applications, however, are likely to be implemented using some combination of the other approaches. The remaining part of this chapter focuses on improvements in the conventional development process and installing an application package, leaving to the next chapter the discussion of newer techniques.

The Continued Use of Procedural Languages

Programs hand crafted in a third-generation procedural language will continue to play an important role in management information systems. Rather than dominating the development process as they do today, however, these conventional languages should be used as a last resort when other approaches prove to be infeasible for some reason.

Machine efficiency, although declining sharply in importance, can still be a design issue under certain circumstances. This is particularly true of a high-volume interactive application that requires the capacity of the largest commercial mainframes. An efficient program may offer the only feasible way of handling large applications of this sort.

In a typical program, a small core of the modules accounts for a large share of the processing time. The 80–20 rule commonly applies: 80% of the computer's time is spent executing 20% of the code. The computer might even spend half of its time within 5% of the code. In an order entry program, for example, a great deal of the computer's time is spent in the small portion that handles the basic functions common to all incoming orders. The parts that handle unusual tasks or prepare occasional reports may require many lines of code but take up a small share of the total processing time.

This skewed distribution has an important managerial implication: it suggests that only a small portion of the code has much effect on a program's use of machine resources. To gain machine efficiency, programmers need to focus on only a small proportion of the total system. For this core portion, programming in a machine-efficient procedural language may make sense. The infrequently executed portions that constitute the vast bulk of the code can be written in a nonprocedural language that emphasizes efficiency in the use of *human* resources rather than *machine* resources. (It should be noted that a program can include modules written in different languages.)

Management of Programming

Since procedural languages will continue to play an important role in systems development work, it behooves an organization to employ the best programming practices available. A sustained productivity gain of perhaps 10% per year is a reasonable—if not monumental—goal for improving conventional programming.

It is difficult to give a succinct description of good current practice in programming. Basically, it comes down to employing sound engineering management techniques in a supportive environment for the programmer.

As in any complex organizational activity, management should make clear-cut assignments of responsibilities for each design and programming task. A disciplined scheduling methodology should be adopted, with each task having a formal written schedule, set of specifications, and resource requirements. Each step of the process should have a well-defined work product, such as a design specification or a working program module. Periodic reviews and formal feedback mechanisms should be established to maintain good coordination among the project members and to keep the schedule up to date. Although the cost of managing a large project is not trivial—it can easily reach 5% of the project budget—skimping on the management function is an expensive economy.

Modular Program Structure

An explicit and carefully managed modular structure is a critical requirement for a system's successful development and eventual maintenance. With such a structure, all information passed between modules takes place through well-defined interfaces. Coordination can thus be limited to managing the interfaces, because any change within a module that does not change an interface will not affect the other modules. The intent is to partially segregate the modules so that their development can proceed in relative isolation, without detailed and continual coordination with other module teams.

Reducing the Size of Project Teams

The structure of individual project teams varies among organizations. One practice that has been used to good effect is to break the staff into small teams, each of which is headed by a **chief programmer**. The chief programmer serves as the master craftsman, with apprenticelike helpers to assist in the less skilled portions of the work.

Individual teams should be kept small to reduce the problems of coordination among team members. The number of pairwise communication links among members grows at a rate greater than the square of the size of the team. For example, a team consisting of three members (A, B, and C) has only three such links (A-B, A-C, and B-C); doubling the size to six persons brings the number of links to 15—five times as many as before. Each link corresponds to the communication and coordination that must be maintained between the pairs of team members. For a task that requires as close coordination as a complex programming project, it is obviously desirable to limit the number of such links.

A team size of one is ideal from the standpoint of coordination, but for a large task the amount of work involved would typically require an unacceptably long elapsed time to complete. An application estimated to take 500 person-days, for example, would require one person for 500 work-days, or nearly two years. If the work content remains constant, a two-person team could finish the task in about a year, and a four-person team in half a year.

Unfortunately, the work content does *not* remain constant with increased team size. As team size grows, the effort that must go into coordination begins to consume an increasing share of the team's time. Eventually the additional coordination time exceeds the productive time of an added team member, and *elapsed* time (not to mention the work content in *man-months*) actually goes up. In his seminal book, *The Mythical Man-Month,* Frederick Brooks points out the hazard of trying to reduce the development time for a project by adding bodies, especially in a crisis situation in which the team faces a looming missed deadline. It is usually far better for management to reconcile itself to a late schedule than it is to add staff in a misconceived attempt to buy time.

A team size of three or four persons often strikes a good compromise between reduced coordination and stretched-out schedules. The members can complement each other's skills, permitting a certain degree of specialization of responsibilities. A team of this size also provides continuity and backup in case a team member leaves the project, and it offers the junior members opportunities for apprenticeship training.

For large projects, involving perhaps several thousands of man-months of work, a team of three or four persons can scarcely make a dent within an acceptable time interval. A long development period would delay the benefits from an application and make it difficult to keep up with the changing user needs that occur during an extended development cycle. Large projects inescapably lead to large teams.

Project size can be reduced somewhat by keeping a system's initial deployment to the minimum workable capabilities, followed later by the development in bite-sized chunks of deferrable add-on functions. A modular program structure also helps, since it allows multiple teams to work on relatively independent parts of a large system with minimum need for coordination across team boundaries. These approaches have definite limits, however, and large projects inevitably pay a heavy price for their size.

Team Continuity over the Life Cycle

An important issue in project management is the continuity of the team over the life cycle of an application. At one extreme, each stage could have its own team participants, with little overlap across stages. This is not uncommon in practice, as a matter of fact, especially for large projects in large organizations. A few senior analysts may handle the feasibility study. Different groups, also generally with a good deal of experience, take over the tasks of needs analysis and design. Programmers with perhaps fairly limited experience then do the coding. At the end of the cycle, left with what many technical people regard as the dregs, junior maintenance programmers move in.

This approach offers the advantage of specialization and the efficient use of scarce technical talent for the higher-skilled and more critical tasks; the remaining tasks can then be assigned to the more numerous and lower-paid team members. Such specialization, though quite common, raises some serious problems. It certainly adds to the difficulties of coordination across the development stages. Perhaps more important, it can reduce the stake that team members have in an application's success. Flaws in the early analysis and design stages may be buried in lengthy design documents; when weaknesses become obvious twelve months later as an application goes on line, the reward system may not provide clear-cut feedback in evaluating the work done in the early stages.

As an alternative, the same team—or at least a few critical members—can remain involved throughout the entire development process, and even into the early part of the maintenance stage when rapid adaptation takes place. This scheme offers a number of important advantages:

1. There is very little ambiguity about incentives and rewards.
2. With continued involvement and a shared stake in an application's success, team members and the client community have an opportunity to build a close rapport.
3. Team members with a detailed knowledge of the system do not leave users in the lurch; instead, they remain available to deal with the residual unresolved problems and changing needs that inevitably occur during the early operational phase of a system.
4. Junior members gain the invaluable experience of living through an entire development cycle and seeing the consequences of their efforts.

5. With the increasing popularity of high-productivity tools, the added cost of using high-priced talent for coding becomes quite acceptable.

Although constraints on staff skills and the practical problems of personnel assignments may reduce the opportunities for a team member's conception-to-birth involvement with an application, an approximation of such commitment seems like a worthy goal.

Increasing the Productivity of the Development Environment

Increasing the productivity of the technical staff offers the obvious benefit of reducing development costs; it also offers the less obvious advantage of permitting smaller teams to complete a given task within a limited elapsed time. Although it is probably true that major productivity gains will come only through basic methodological changes (as represented by fourth-generation tools), it is nonetheless true that some worthwhile improvements can also be made in the use of conventional procedural languages. These languages will continue for a number of years to soak up an important fraction of the development effort in most organizations, and so even a modest increase in productivity carries a high payoff.

The environment within which programming takes place strongly affects productivity. The following characteristics are common features of a productive environment:

- Interactive program development, with subsecond response times for most tasks not requiring heavy computing (such as editing the program for syntax errors and performing small test runs).
- Interactive **debugger** program that provides diagnostic aids to assist the programmer in locating and correcting bugs in the source code.
- Text editors that provide features tailored to program development—for example, powerful word processing capabilities for duplicating and modifying code, multilevel indentation of the

code to show its hierarchical structure, built-in syntax checking, and fast switching from typing the code to running it.
- A capability for splitting the terminal screen into logically separate **windows** so that the programmer can, for example, assemble program fragments from several different windows, or type new code in one window and simultaneously examine the resulting output in another window.
- A **program library** with powerful retrieval capabilities so that a programmer can assemble a new application program out of portions of existing code (thus avoiding duplication of effort in programming the common functions that occur throughout most applications programs).
- Testing facilities that simplify the task of generating test data, running test programs, and analyzing the results.
- Electronic mail that permits a team of designers and programmers to work cooperatively on a tightly coordinated basis.

Through such means, useful progress is being made in the creation of a more productive environment for software development. Most companies have probably under-invested in the development environment. It would pay these organizations to devote a larger portion of their budget to the provision of a more productive environment for their third-generation programmers. The fruits of such an investment will in many cases also carry over to the fourth-generation environment.

Human Resources

Finally, not to be overlooked in the search for improvement in the software development process is the quality of the technical staff. The productivity of analysts and programmers can differ enormously—by a factor of ten or more—due to their innate ability, training, and motivation. The organization must focus its efforts on increasing the proportion of the highly productive members.

An organization with a capable staff benefits from a strong positive feedback process: good people can select and attract other good people. They welcome opportunities to keep up with important new areas of technology and to work on interesting projects. Contrary to a common stereotype concerning computer people, effective staff members want more than just a technical challenge; they also want to contribute to the strategic goals of the organization.

This suggests that the selection, training, and management of technical staff may be an overriding determinant of success. With a high-quality, well-motivated staff, almost any methodology works; without it, almost nothing does.

Use of Purchased Application Packages

An organization can avoid some of the problems of developing its own custom-made application software by installing an existing commercial package. The market now offers an increasingly large variety of products that meet a wide range of needs—from generalized programs (such as payroll or general ledger) to specialized niche products (such as a program for financial analysts dealing with mergers and acquisitions). In the vast majority of cases, these products are developed and marketed by **third-party vendors** rather than the producers of the hardware on which the programs run.

Pricing of Software Products

The license fee for a software product can be considerable—as much as $250,000 or more in the case of some mainframe products. If the product is installed at multiple sites (or even on different hardware at a common site), the price generally goes up but with a declining fee for each machine. It is becoming quite common for vendors to set the price on a "value added" basis—dependent on the maximum number of users supported on an interactive computer, for example. The initial license fee sometimes grants perpetual rights to use the software and sometimes grants the right for only a fixed period of time. In any case, the vendor almost always retains full ownership of the product.

Micro-based products tend to have a much lower unit price. Even here, though, software fees can swell as these products get adopted widely throughout the organization. The fee for a hundred copies of a product with a unit price of $500, for example, jumps into the mainframe price range. To contain these costs, many using organizations actively pursue fixed-fee site licenses or steep discounts for multiple copies of software.

In addition to the initial license fee, a user may also have to pay an annual maintenance fee. This generally entitles the user to periodic updates and enhancements to the product. Mainframe and minicomputer products almost always call for an annual fee, typically in the range of 10 to 15% of the initial installation fee. Most micro products do not carry such fees (except for optional periodic program updates).

Selecting a Product

Selecting a suitable package is a nontrivial exercise. There are literally thousands of products on the market, covering virtually all types of hardware and applications. For the more popular applications, dozens of packages may seemingly meet the needs of a given user. Choosing the best product, or even an acceptable one, is an expensive and time-consuming process.

Some publications and consultants specialize in keeping track of available software products, and almost every package vendor can present impressive glossy brochures. These can help considerably in narrowing the choice, but the final selection from the short list of plausible products is still difficult, time consuming, and expensive. And no matter how careful the selection process, there is always the risk that the chosen product will prove to be unsatisfactory in practice.

Selection of a package does not eliminate the need for a careful needs analysis. On the contrary, a considered set of needs provides the only intelligent basis for choosing among competing products. Needs should be classified by importance, equivalent to the "must do," "should do," and "nice to do" categories discussed earlier (although some buyers use a two-level classification, such as "mandatory" and "desirable").

With a well-defined set of needs in hand, the customer can rank the various packages according to how well they meet the company's needs. A numerical weighting scheme is sometimes used for this purpose. Each need is assigned a weight, and then each product is scored with respect to its degree of satisfying each of the needs. The overall ranking of a product equals the sum of its weighted scores across all needs.

In the preliminary screening of products, one must rely primarily on the vendors' technical specifications. For products that pass this test, comments from existing users provide invaluable insights into the quality of a product and its support from the vendor. Ultimately, though, a product should be subjected to a detailed hands-on evaluation before a final commitment is made. There is no substitute for trying before buying, because until one actually uses a product it is very difficult to assess such matters as its ease of learning and ease of use, the power and breadth of its features, its response time, and its freedom from bugs. In the case of a large-scale package where installation costs are substantial, an on-site trial is generally not practical. (On some large contracts—particularly for government agencies—selected software vendors may be required to prepare **benchmark programs** to demonstrate that their products can successfully handle the intended application).

Adapting a Program Package to Specific Needs

A successful product typically offers a variety of functional options, such as a payroll program that allows several different pay schemes (hourly, weekly, or monthly rates, say), deductions (taxes, union dues, savings clubs, insurance, etc.), and taxing jurisdictions (e.g., municipal, county, state/provincial, and national). In addition to its tailored functional capabilities, a good package also allows considerable generality in specifying things like the content and format of the user interface.

A package provides its generality through a set of user-defined **parameters** that govern the program's execution. To include a municipal tax in a payroll program, for example, the user could specify the option and the withholding rate by setting a designated parameter to the appropriate value—parameter 25, say, set equal to .03 to effect a 3% withholding rate; the identification of the deduction on the printed pay stub (e.g., "Philadelphia Wage Tax") could also be designated by the user. Similarly, in an accounting program, the user can generally specify the names and classification codes of the chart of accounts. The parameters needed to tailor the program to a given customer are typically stored in the form of an accessible table. If any of the parameters in the table are modified, the program executes according to the new set of values.

Tailoring a package through parameter specification requires that the vendor anticipate the range of options desired by the customer population to which the product is directed. These options must then be built into the product. There is a definite limit to such generality, because each option adds to the cost of designing, programming, and supporting the product.

Any required variations that cannot be met with the existing set of built-in options requires a modification to the actual program code. This may be relatively easy or quite difficult, depending on the nature of the change and the way the package is designed and supported. If the change is fairly straightforward, if the package has a well-defined modular structure, if the user is furnished the source code for the package (or if the vendor is willing to contract to make the change), and if the source code is well documented to aid the programmer making the change, then modification may present no great problems. If, on the other hand, one or more of these conditions is not satisfied, the change may range from difficult to impossible. In that case the customer has the choice of abandoning the product, incurring a high cost of making the change, or forgoing the benefits of having a system closely tailored to its specific needs.

Advantages of a Package

A software product can offer a number of advantages under the right conditions. The cost of buying a package is generally considerably less than implementing a custom-designed system, because the software vendor can amortize the development cost over multiple customers. This is particularly true of most software for personal computers, where a vendor may anticipate potential sales in the hundreds or thousands of units. As a result, some micro-based application packages priced at a few hundred dollars may match or exceed the functional capabilities of mainframe products selling for $50,000 or more.

A package can save installation time as well as money, thus providing earlier returns on the software investment. It can also reduce technical risks, since a package is generally based on an existing proven system. An organization that wishes to avoid the difficulties of hiring and managing its own technical staff may rely instead on an outside vendor to install and maintain a turnkey system.

The quality of the best packages is often very high. With the potential for multiple sales, an established vendor can afford a heavy investment in a product's development and support. A high-quality commercial product may cost the vendor perhaps ten times the amount that it would cost to develop a working program for a single user. These added costs show up in such forms as a wider variety of functions included in the package, a more user-friendly human interface, greater reliability achieved through exhaustive debugging and continual correction of bugs reported by users, better documentation and training materials, and better customer support.

A software package can provide an effective mechanism for technology transfer and organizational learning. The product itself may embed a great deal of technology related to a given application area; this may be further augmented by the knowledge of the vendor's support staff. An inventory control package, for example, may include advanced statistical techniques for determining order points and order quantities. Even a routine product like a general ledger program for a small business may incorporate knowledge of accounting procedures that would be difficult for the proprietor to acquire.

There are risks, of course, in installing a "black box" product that the using organization does not understand, or in slavishly adhering to the forms and procedures imposed by the product. To minimize these risks, management should assign internal responsibility for understanding the product and adapting it, where necessary, to the special needs of the organization (or adapting the organization to the product, which sometimes provides the better tradeoff, especially for small organizations).

Disadvantages of a Package

For all their advantages, application packages also have some important limitations. As we have seen, selecting a satisfactory package is not easy. This is not a minor issue: in some cases the effort merely to find a suitable product adds significantly to its total installation cost.

We have also seen that installation and maintenance fees can be quite high. Although these costs may compare extremely favorably with a third-generation custom program, they look much less competitive in comparison with a system developed with fourth-generation tools. Under certain circumstances, the best choice may swing away from a package to a custom-designed system based on a 4GL.

The economy of a package hinges significantly on the added cost of adapting it to specific user needs and integrating it with the organization's existing system. If a package can be installed successfully with little or no changes, and has a simple interface with other parts of the system, then it is very difficult to match the advantages of the package. As the number and complexity of the changes and interfaces grow, however, the package begins to lose its competitive advantage. In complex installations, it is not at all uncommon for the cost of adaptation to exceed the initial license fee by a significant amount. Eventually the point is reached where it is cheaper to develop a custom program than to make extensive modifications to a package to shoehorn it in where it does not fit very well. This crossover point is reached all the earlier as the cost of custom programming drops with the use of high-productivity development tools.

Modifications to a package not only add to its cost, but also to the technical risk. Changes to any program are risky, and a package is no exception. This is particularly true if the vendor makes no special effort to facilitate program modifications. When the client's programmers make the changes, they inherit responsibility for maintaining a program written by the software vendor. Management can reduce this particular risk by contracting with the vendor to make the modifications, but this increases another type of risk—long-term dependency on an external party.

Most successful packages have been around in the marketplace for several years. They tend to be based on third-generation technology. Such a package may meet current needs quite satisfactorily, but it will probably exhibit many weaknesses of third-generation systems with respect to its ability to adapt to changing needs. Even if a package is based on a fourth-generation language, that 4GL may not be the one on which the client wishes to base the rest of its MIS.

Finally, owing to its widespread availability, an off-the-shelf package is not likely to confer a significant competitive advantage on any of its users (although some organizations may use the package more effectively than others). If an organization seeks a strategic advantage from information technology, it will almost surely have to develop its own custom software that differentiates itself in some important way.

Summary Comments on the Use of Packages

Most organizations of any size will probably find that a mixture of packages and custom-designed programs provides the most cost-effective information system. Purchased products will very likely claim an increasing fraction of their software budget. Flexible application packages will retain an important share of the market, but many of the successful products will aim at providing users with high-productivity tools for implementing their own custom-designed systems.

The ideal use of application packages seems to fall into the following categories:

- A set of well-integrated applications for a relatively small organization willing to accept the programs without making any changes (but often requiring some modifications in the user's policies and procedures to adapt to the packages); included in a typical set of applications (usually developed, perforce, by a single vendor) might be general ledger, order entry, inventory control, accounts payable, accounts receivable, and payroll.
- An application peripheral to an organization's mainline activities that has a well-defined and relatively simple interface with other parts of the MIS (such as a payroll program or a system for stockholder accounting and record keeping).
- An application program that embeds specialized and proprietary expertise and interfaces relatively cleanly with other parts of the MIS; a product for providing computer-assisted financial advising for medium-income clients of a financial services firm would probably fit this description.
- A complex application in an area in which the company does not feel that it can gain a significant competitive advantage, such as a manufacturing planning system in a marketing-oriented firm.

These categories are certainly not clear-cut; the extent to which a given application meets the criteria is largely a matter of degree. It is fair to say, though, that a package becomes less appropriate as its functions move closer to the heart of the enterprise. Capabilities on which the organization aspires to build a strategic advantage almost certainly must come from custom-developed programs. To achieve a sustainable success, these programs must undergo a continual process of adaptation as the organization seeks to stay ahead of the competition and exploit newly revealed opportunities. This places a high value on an organization's ability to master the use of high-productivity development tools.

Further Readings

Brooks, Frederick P., *The Mythical Man-Month,* Addison-Wesley, 1979. A classic in the systems development literature, dealing with the problems of implementing large-scale systems.

Gane, Chris and Trish Sarson, *Structured Systems Analysis,* McDonnell Douglas Professional Services Company, 1985. Good coverage of a formal procedure for MIS design.

Jackson, Michael, *System Development,* Prentice-Hall, 1983. Similar in coverage to the Gane and Sarson book, it discusses a formal design methodology developed by the author.

❖ 7 ❖

High-Productivity Development Tools

Fourth-Generation Languages

The word breakthrough gets used to excess in describing technological advances. In reality, progress is much more likely to come from cumulative improvements over an extended period of time than from a sudden dramatic new development. So it has been in the case of software development tools. Fourth-generation tools provide a continuation of the process that has been going on since the first high-level programming languages: moving away from the constraints of machine architecture toward languages humans find more congenial for describing computational tasks.

We know from past experience that a new language generation takes a considerable time for assimilation. Information systems and practitioners exhibit a great deal of inertia. Organizations have an enormous investment in the existing software base and the skills of analysts and programmers. It takes a long time to disseminate the latest technology, train practitioners in its use, and incorporate it in working systems. The process is slow enough when an organization has adequate resources and a clear vision of where it wants to go and how it will get there; it is all the slower in the real world of skimpy discretionary resources, uncertain goals, and a variety of competing technologies.

Despite these difficulties, a few leading organizations have made major strides toward applying fourth-generation technology. Most other organizations have at least begun serious efforts to move in this direction. Laggards will still be struggling with obsolete technology for a number of years, and will probably pay an increasingly heavy price for their sluggishness.

Advantages of 4GLs

Although the process of assimilating fourth-generation technology is slow, it is nevertheless inexorable. Few developments in the field of information technology will have an equivalent impact on the way we design and implement information systems. The new tools offer a number of extremely important advantages that no organization can afford to ignore:

- They lower the cost of program development, potentially by as much as a factor of ten or perhaps even more.
- Their increased programming efficiency makes it feasible to develop inexpensive prototype systems.
- They reduce the work content of programming, permitting small teams to develop major application programs within an acceptable elapsed time and without many of the serious coordination and quality control problems associated with large-scale projects.
- The smaller size of a nonprocedural program generally results in a corresponding reduction in the effort to test and maintain the code.
- The relative ease of learning and using 4GLs permits nontechnical end users to develop some of their own applications.
- A fourth-generation system provides an enabling technological base for building an adaptive, responsive, and cost-effective strategic information system.

What Is a Fourth-Generation Language?

Chapter 3 introduced some of the general characteristics of fourth-generation languages; we now need to amplify some of these ideas. There is little consensus as to a precise definition of fourth-generation languages. A much clearer consensus exists as to what is *not* a 4GL: the widely used procedural languages—COBOL, Fortran, BASIC, Pascal, C, and Ada—are certainly not.

A common characteristic of languages generally regarded as fourth generation is their nonprocedural nature—the specification of *what* is to be done rather than *how* it is to be accomplished. A query language, for example, might specify the following task:

```
SELECT CUSTOMER, STATE, CODE, AMOUNT
FROM CUSTOMER_HISTORY
WHERE REGION=SW
    AND (CODE=1412 OR CODE=1413)
    AND AMOUNT>10000
ORDER BY AMOUNT DESCENDING
```

This would be translated automatically into a machine language program that generates the specified report.

An equivalent COBOL program, say, would take perhaps ten times as many lines of code as the query language. Empirical studies suggest that a programmer writes about the same number of lines of code per day, regardless of the particular language used. On this basis, the COBOL program would thus take about ten times as long to write and debug as the inquiry language. The conciseness of a nonprocedural task definition accounts for much of the advantage that a 4GL enjoys over a procedural language.

Language Translators

A nonprocedural task specification implies that the interpreter of the language can translate the desired end results into an equivalent procedural specification (i.e., into the computer's machine language, which is thoroughly procedural). In the above inquiry, for example, the language translator interprets the expression ORDER BY AMOUNT DESCENDING to mean that the customer records passing the various selection criteria (customer located in the Southwest region, etc.) must be sorted and printed in descending order by the amount of sales to the customer. In other words, the translator must have built-in knowledge about the subject matter described by the nonprocedural language. The more a language relieves the programmer from having to deal with detailed procedural matters, the more knowledge has to be built into the translator.

The requirement for built-in knowledge generally limits the range of problems for which a 4GL is ideally suited—after all, the translator cannot know a lot about a wide range of problem contexts. The **spreadsheet language** Lotus 1-2-3, for example, knows about the calculation of a net present value when given a cash stream and a discount rate; it also knows how to determine the number of days between two specified dates. It is thus very effective in dealing with many financial planning problems involving discounted cash flows and time periods. It would not, however, be suitable for programming a simulator of a supermarket checkout operation; for that kind of problem, one would use a language that incorporates knowledge about such matters as random arrivals, management of queues, and service times.

User-Friendly Language Features

Lotus 1-2-3 is an example of a nonprocedural language designed to facilitate problem solving by a nonprogrammer user. To be successful in this role, skills in the language must be relatively easy to acquire and easy to retain with only casual or infrequent use. Unless a workable core of a language can be learned in two or three days, it is unlikely to capture the interest of the nonprogrammer professional. For many busy senior executives, learning time may more likely have an upper limit of an hour or so.

A 4GL—particularly one designed primarily for the nonprogrammer—must aim at making the language as **friendly** as possible. Friendliness is in the eye of the beholder, to be sure, but the following features are common in languages regarded as friendly:

- Inclusion of high-level built-in functions—language **primitives**—designed to aid in performing tasks found frequently in the problem context for which the language is targeted, such as the calculation of net present value and days between calendar dates in a language designed for financial analysis.
- An easy to remember and familiar terminology, such as "NPV" to designate the function that calculates net present value.
- A familiar language **syntax**, such as the use of common algebraic notation in a language designed for scientific and engineering applications.
- Familiar user interface, as in a spreadsheet language that displays information in a tabular format common in many business contexts.
- Provision for easily linking the output of one part of a computation with other related parts, either by incorporating all parts within a single integrated package (e.g., spreadsheet calculations, data management, graphics, and text processing) or by employing a standard interface to facilitate the **seamless** exchange of information among separate, more specialized languages.
- A consistent and intuitive language that presents few "surprises," allowing the user generally to predict what the language will do in a given situation, guided by a general understanding of its logic and syntax rather than a recollection of its detailed features.
- Intelligent **default** decisions on the part of the system so that options not specified by the user will generally lead to reasonable outcomes—for example, in the absence of an explicit specification of column headings and widths, the system will format a report in a way that makes it readable and attractive.
- Functional subsets that permit a user to take advantage of parts of the language without knowing all of its features (and without getting into trouble)—for example, allowing a manager to spec-

ify a simple ad hoc report without knowing how to use the arcane programming features of a query language.
- A depth of capabilities, so that users can tackle increasingly complex problems as their knowledge of the language grows.
- Incorporation of a useful set of low-level primitives that can be combined to perform a more complex task not defined as a built-in function of the language.
- A language that permits the user easily to extend its capabilities by defining a tailored higher-level function in terms of existing built-in functions; once extended in this way, the added function can then be used as an integral part of the language (including its use to define other functions).
- An undemanding language that does not impose unnecessary burdens or constraints on the user, such as rigid punctuation or formatting requirements.
- A relatively safe environment that tries to protect the user from making serious errors (e.g., by warning the user before deleting a file).
- A forgiving environment that does not extract a large penalty if the user makes a minor error—(e.g., the provision of an UNDO key that reverses the effects of the previously invoked operation).
- An interactive environment that provides fast feedback from a computation so that errors can be quickly located and corrected.
- Insightful computer-generated diagnostics to detect (and even to correct in some cases) errors in a program.
- An on-line HELP facility that permits the user to request specific information about the language.
- A productive environment for editing and revising a program, such as a **full-screen editor** that allows the user to move to any part of a CRT screen to make additions or deletions to selected parts of the program.

Each of these features exists to some degree in fourth-generation languages (and even earlier generations), but no 4GL can claim strength in all the features. The trend is certainly toward continued improvements and extensions to make the 4GLs more productive problem-solving tools.

Languages for Developing Decision Support Systems

Languages designed especially for developing decision support models illustrate very well the general principles of fourth-generation languages. Some of these languages have been around for a number of years, initially for mainframe machines and now increasingly for micros. Many of the more popular languages, such as IFPS from Execucom, have versions for both ends of the hardware spectrum. This allows the user to develop a model on a micro and then execute it on a mainframe, or even to divide a common computing task between the two (the mainframe managing a central database and performing heavy-duty computing, for example, and the micro manipulating extracted subsets of the data and graphically displaying the results).

Spreadsheet languages—especially Lotus 1-2-3—now dominate the modeling scene. They have already taken over most of the low-end applications. As their power has grown in term of functional capabilities and model size, and as the population of experienced users expands, they find use in ever more ambitious applications. It is often the case, however, that a user eventually wants to tackle problems that exceed the limits in size or complexity of the models that can conveniently be handled with a spreadsheet language. At this point the user needs the features, data storage capacity, and computing power of a more powerful modeling language on a larger machine.

All of the DSS languages provide powerful computational capabilities. Algebraic relationships can usually be defined in terms of symbolic variable names, such as GROSS MARGIN = REVENUE − COST OF GOODS SOLD (although in spreadsheet languages a variable is usually identified by its physical position on the spreadsheet— i.e., its column and row coordinates). Other more complex relationships can be defined—for example, involving exponential and logarithmic functions, useful in dealing with compound rates of growth. Statistical computations, such as calculating the mean and standard deviation of a sample, are generally built in as primitives of the language.

All modeling languages incorporate some of the common

business functions, such as net present value and internal rate of return. Other popular built-in functions are calendar calculations and a variety of depreciation methods (e.g., straight line or declining balance). In choosing which functions to include as primitives (and hence, by exclusion, what the user must otherwise program in more detailed form), designers of a modeling language must trade off the perceived convenience of being able to invoke a rich variety of powerful primitives versus the added cost and complexity of learning to use a comprehensive language. Many of the fancy functions that look good in sales literature may seldom be used, and yet they generally impose a psychological hurdle on learning the language.

Any model of even modest complexity requires the specification of conditional relationships that define the value of a variable based on the value of other variables. In a budgeting model, for instance, "fixed" administrative costs may increase in step fashion with an increase in the aggregate volume of sales. Such a relationship could be defined in Lotus 1-2-3 by the statement @IF(SALES < 100000, 8000, 11000), which sets administrative costs at $8000 if sales are less than $100,000, and at $11,000 otherwise.

Most modeling languages provide features that assist the user in performing sensitivity ("What if . . .") analyses. For example, the user may be able to define a range of values for one variable, and the system will then compute the corresponding values of all other variables affected by the changes. Sensitivity analyses are an essential part of using a decision support model, and so facilitating them adds real value to the language. Some languages have a **goal-seeking** feature that allows the user to define a desired end result ($2.50 per share earnings, say), and then the computer will find the value of a specified related variable that will achieve the stated goal (e.g., sales revenue must equal $3.7 million to achieve $2.50 per share).

Vendors of modeling languages compete on the basis of features they offer. Representative of some of the more powerful features are the following:

- Database management functions that allow the user to extract and aggregate data from a shared database at the corporate or divisional level.

- Report formatting for the preparation of custom-tailored and attractive reports.
- Preparation of graphic displays based on the outputs of a model.
- Integration of text processing with model outputs.
- Consolidation of lower-level models (by P&L center, say) into a higher-level aggregate (e.g., a corporate-wide report).
- Optimizing algorithms.

Use of Fourth-Generation Tools for Transaction Processing Systems

Languages used for DSS illustrate the general principles that motivate the use of fourth-generation tools by nontechnical end users. Most of these principles also apply to transaction processing systems developed by professional programmers. In both cases the intent of the language is to provide a productive set of tools and development environment for building application software.

Although applications differ widely within the same organization and across different companies and industries, they tend to include a number of common functions. A 4GL for a transaction processing environment should take over most of the burden for implementing these common functions, and provide productive tools for handling the idiosyncratic tasks. For the dominant system architecture evolving in the 1980s and beyond—interactive distributed systems to support operational activities throughout the organization—the following functions reappear in almost every design:

- Management of a large database that involves complex relationships among its data elements (such as the links of an inventory item with its supplier and with the customers who have bought it).
- Preparation of a variety of reports, both standard periodic reports and ad hoc queries.
- Specification of interactive screen formats and the editing requirements for input data.
- Managing the user input interface, through such means as menu selection, user commands, or question & answer dialogue.

- Managing communications from a variety of sources throughout the network.
- Managing the flow of a transactions through all of its processing steps—for example, logging its entry into the system, giving it an appropriate priority, putting it into queues and scheduling its processing, and generating a journal to provide an audit trail of the transaction.
- Providing backup and recovery if a failure occurs.

A development language that handles these functions efficiently can go a long way toward achieving the tenfold increase in programmer productivity claimed by advocates of fourth-generation tools. Consider, for example, the task of programming screens for an interactive application. This is just one of many tasks involved in developing a new application, but it illustrates very well the potential productivity gains that can come from the use of a 4GL. Figure 7-1 shows a typical screen.

Programming this screen in COBOL, say, would require a considerable effort. The program would have to create a complete image of the screen in the computer's memory. It would have to differentiate between the field names (e.g., CUSTOMER NUMBER), which remain unchanged for all customer orders, from field values (002537), which are entered from the keyboard by the data entry clerk for each new order. Various editing checks would have to be programmed to determine if the entered data fall within the prescribed range of values. A large application program may require hundreds of such screens, adding significantly to development time.

Implementing a screen format becomes almost a trivial task with the use of a 4GL screen formatter. (Choosing the fields to include on the screen, and defining their editing specifications, involve difficult design questions, but that's a different issue.) The programmer can typically "paint" the form on the screen by typing in the name of the fields and the required spacing for the variable field values. Defining editing specifications is generally a matter of filling out a table that gives the set of permitted values or their ranges. Default values can be established for any field to save keyboard entry time, such as automatically filling in today's date on a customer order unless the

CUSTOMER ORDER FORM

CUSTOMER NUMBER [002537] DATE [11–22–87]

CUSTOMER NAME [HENRY A. FERGUSON]

STREET ADDRESS, LINE 1 [APT 3B]

LINE 2 [2307 PINE STREET]

CITY [PHILADELPHIA] STATE [PA] ZIP [19103]

PAYMENT METHOD [1] 1 CREDIT CARD 2 CHECK 3 CASH

CREDIT CARD [3] 1 VISA 2 MC 3 AE 4 DC

EXPIRES [7–88]

CATALOG NUMBER [1242]

ITEM NUMBER	ITEM NAME	SIZE	QUAN	UNIT PRICE	AMOUNT
87393	GE TOASTER		1	32.50	32.50
84762	STOCKINGS	11	4	3.25	13.00
					45.50

Figure 7-1. Order entry screen.

clerk explicitly enters another date. Changes in the screen format can generally be made quite easily—for example, by using a keyboard **cursor** or a mouse to drag a field to a desired new location on the screen.

A screen formatter is quite typical of the high-productivity tools incorporated in 4GLs. Programming reports and handling user inputs from a terminal provide similar economies when handled with a 4GL. These tasks can be excruciatingly time consuming when programmed in a procedural language, but become relatively simple with the appropriate set of 4GL tools.

Use of a 4GL continues to pay dividends after the initial development. Since fewer lines of code need to be written, testing is proportionately easier. Not only is it easier to correct program bugs, but errors in the design itself can be corrected much more easily because of the simplicity and power of the language. This advantage holds during the program's maintenance period, which typically contributes most of the life-cycle software costs.

As powerful as they have become, 4GL development tools are not appropriate for all programming problems. In a large, complex application there will almost always be parts of the program that cannot be handled satisfactorily with the high-level primitives of a 4GL. Its standard language features, for example, may not handle the complex program logic required of some applications. The relatively inefficient use of hardware resources by the 4GL might add substantially to capacity costs (or, in the case of an exceedingly high-volume transaction processing system, even render an application infeasible). Under these circumstances, it is essential to provide the programmer a means of gaining easy access to the generality and machine efficiency of a procedural language.

A built-in procedural language is becoming a common feature of 4GLs designed for heavy-duty system development work. At the microcomputer level, for example, DBase III has an integrated procedural language equivalent in power to such languages as BASIC and Pascal. Because the procedural capabilities are an integral part of the language, passing back and forth between procedural and nonprocedural programming is made very simple.

The more powerful mainframe-oriented products have similar capabilities. In cases where the built-in procedural language facilities are not adequate to a given programming task, most 4GLs provide "hooks" to interface with conventional languages like COBOL. This gives the programmer the option of exiting from the 4GL, performing a task in the procedural language, and then returning to the 4GL with the computed results. This would permit efficient programming of the heavily used core of an on-line system, for example. Since most applications consist largely of undemanding tasks that can be handled well in the 4GL, the fraction of the program that must be written in a procedural language is generally fairly small—perhaps 20% at most, and often much less.

Database Management Systems

Data management has a central role in a fourth-generation development environment. Effective management of data resources is increasingly recognized as one of the critical enabling capabilities needed to develop a successful MIS. Data management includes a number of functions connected with the storage, retrieval, and protection of the organization's data resources. These functions are provided by means of a system software product called a **database management system**, or **DBMS**.

Relationships Among Records

A critical function of data management is defining relationships among stored data elements. The database for a distributor of industrial products, for example, would typically include records for customers, inventory items, suppliers, customer orders, and purchase orders (shown earlier in Figure 4-1). Figure 7-2 illustrates some of the important interrecord relationships for such an application.

Consider the case of the links among customer records. The need to process a set of these records might arise in a number of ways. Suppose that a user asks for a report analyzing customer purchases. Such a report would present no conceptual difficulty in a conventional system. The customer records would typically be stored sequentially on magnetic disk or tape, and so generating the report merely requires accessing the records in their physical sequence and performing the required processing. So far, so good.

As we saw in Chapter 4, however, a problem arises when a transaction or report requires access to more than one record type, which is certainly not an infrequent need. Suppose, for example, a customer calls to find out the status of an outstanding order. Customer C in Figure 7-2 has two such orders, X1 and Z2. If these are stored in a separate customer order file,

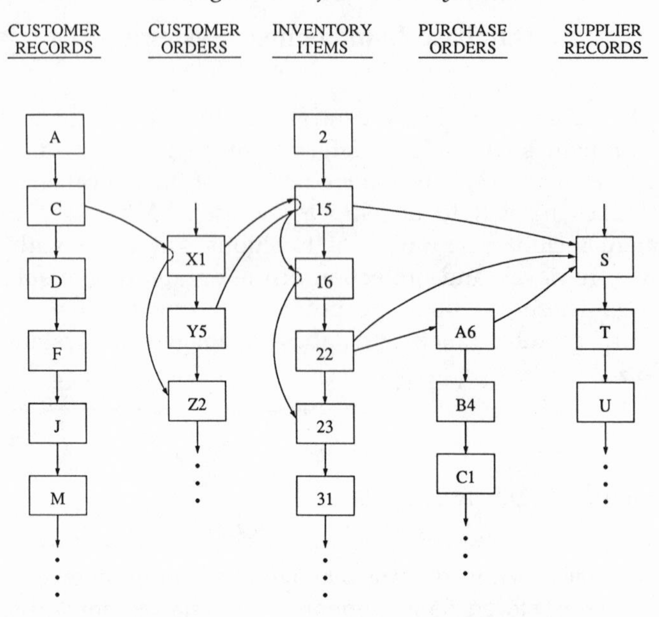

Figure 7-2. Links among database records.

processing the customer's inquiry then requires some way of linking customers to their orders. One way to avoid the problem would be to store the order information as an integral part of Customer C's record, but that has its own disadvantages (such as making it more difficult to obtain access to customer order data alone).

The problem gets worse when we consider the relationship of inventory items to customer orders. An inventory item is not uniquely associated with a specific customer, and so inventory data cannot be stored as part of a customer record. We are forced, then, to deal with interrecord relationships. A similar need arises when we want to relate an inventory item to its outstanding purchase orders and, in turn, to the supplier to which each order is issued.

An MIS that aspires to provide responses to any reasonable query cannot avoid the need to deal with a rich variety of links among data elements. Some information needs can be satisfied very nicely using a conventional structure of homogeneous files, but many other needs would be impractical to handle without an advanced data management capability. Consider the following queries, for example:

1. What is last quarter's sales volume from all customers having unshipped orders older than two months?
2. Which customers have Item 23 on order?
3. Which customers have unshipped orders for items supplied by Supplier S?
4. Among our customers buying more than $10,000 over the last quarter, who bought a product obtained from a supplier with a quality rating of less than the average rating for all suppliers?

Each of these queries could arise from a perfectly plausible business need, and each is answerable on the basis of the information stored in the database. But without a technology that handles interrecord links, one would have to manually dig through and select the information from voluminous computer-generated reports. An important function of the more powerful database management systems is to provide this kind of retrieval capability. (It should be noted, however, that the above queries would not be particularly easy to specify in most existing query languages—certainly beyond the skill of the average casual user.)

Linking records calls for no magic: *logical* relationships among data elements—those governed by the underlying meaning of the data—must be realized in some *physical* form that allows the computer to associate related data. There are two basic ways of doing this, however complex the relationships might be.

In a **network** or **hierarchical** DBMS, a linkage may be established through a **pointer address**. For example, in Figure 7-2, customer record C could include the physical address of order X1; a similar link could also be established from X1 to Z2. The system can thus retrieve the two unshipped orders for Customer C by following the chain of linked records. A customer order, such as X1, can be further linked through pointers to the items on the order (Items 15, 16, and 23). In addition, separate backward links can be defined to establish two-way relationships among records. For example, to respond easily to the queries listed earlier, it would be necessary to relate an inventory item to its customer orders and then to its customers.

In a **relational** DBMS, physical links are established through a common identifier in related records. In customer orders X1 and Z2, for example, the customer number for Customer C is stored. Through this connection the system can then retrieve customer information related to an unshipped order.

Although the creation and processing of linked records involve some highly complex technology, most of the complexity is hidden from the user. For example, the DBMS automatically allocates memory and handles the physical links among records. In some systems the user has to specify the connecting path between records, but more sophisticated systems remove this burden when no ambiguity exists.

The power and convenience of interlinked records come, as always, at a price. Links require stored data in the form of pointer addresses or relational identifiers. In a richly interconnected database, these links can eat up half the storage capacity of a disk file. Furthermore, tracing down linked paths can take a great deal of processing and disk access time. Designers thus face a difficult tradeoff between the cost and the value of information. As the cost of information processing continues to plummet, the tradeoff tips in favor of providing more powerful information retrieval capabilities—but never to the point that users should be indiscriminate in their requests for information.

Common Data Definitions and the Data Dictionary

To have data resources serve as the central focus in the MIS design, the organization must evolve toward a standard definition of each data element. Sharing of data across applications and subunits demands such standardization. This is by no means easy to achieve, from both a technical and a managerial standpoint.

In an era of fragmented systems, data consistency was less of an issue. Even now it is not uncommon to find the same entity defined in quite different ways. The sales department may use product codes that differ from manufacturing's codes, and the international division may have still a third scheme. The shipping department may define a sale as occuring when a product is delivered to a common carrier, while the accounting department may use the mailing date of the invoice. The MIS of an acquired subsidiary may have little resemblance to that of the parent company.

Such inconsistencies are problematical enough in a fragmented system, but they are intolerable in an integrated system that tries to tie applications together through data sharing. Although most of the problems of adopting standard data descriptions involve primarily managerial and political issues, data standardization certainly has some important technical implications as well.

To facilitate data sharing, developers of a system need a common depository of information about the content and structure of the database. This is the purpose of the **data dictionary**. The data dictionary includes an entry for each component of the database, which gives the following types of information:

- Component name—for example, CUSTOMER as the name of the customer record, and CUSTOMER_NUMBER as the name of a data element that identifies a particular customer.
- Constituent parts of a record or other collection of data elements—e.g., CUSTOMER = CUSTOMER_NUMBER, NAME, ADDRESS.
- The format of each data element—for example, the format code NNN-NN-NNNN gives the format of a Social Security number, where N signifies a numeric digit, while the format A20 for an employee's last name signifies that the name consists of up to 20 alphabetic characters; this information tells the DBMS how much space to reserve in storage for these data elements, as well as allowing the system to detect a format error (e.g., to determine that 22-45-3523 and W22-45-3523 are not legitimate SSNs).
- The set of allowed values for a data element—e.g., an order

quantity cannot exceed 50 units, or an employee's gender designation is limited to a value of M or F.

- Security specification: who is allowed to read or change each component and who has "ownership" of the component (and hence has the authority to grant or deny read/write privileges to others).
- Sources and uses of each component: where does a component come from (e.g., as entered data, or as the output of a specified program) and what use is made of it (e.g., input to a program or a report).
- A detailed textual definition of the component—for example, INVOICE_DATE is the date the invoice is presented to the delivery agent.

The data dictionary may have either a *passive* or *active* relation with the DBMS. With a **passive** system, no direct link exists with the DBMS. The data dictionary is designed only to serve its human users—designers, programmers, and the **database administrator** (the person who establishes and administers data standards). A passive dictionary is processed by a software product separate from the DBMS, and may even come from an entirely different software vendor. The software typically provides interactive access to the dictionary, along with functions that make it easy to retrieve information about the database and modify its descriptions.

An **active** data dictionary serves these same objectives, but in addition has an explicit and automatic link with all programs that use the services of the DBMS. Application programs access the database via the data dictionary (either when they are translated into object code or at the time of execution), and so they are kept consistent with the current data definitions. An active data dictionary offers some very important advantages, and this clearly is the way DBMS technology is moving.

Data Sharing

Standard data definitions managed through the data dictionary greatly facilitate the sharing of common data. This can yield substantial payoffs. Suppose, for example, that several applications use employee address data (for payroll, personnel functions, mailing of the company magazine, etc.). Rather than making each application responsible for collecting and maintaining employee addresses, a single authoritative version (coming from payroll, say) can be shared among all applications. This reduces the redundancy of the database, and thus saves data collection and storage costs. More important, it provides a consistency across applications that would otherwise be impossible to obtain. Sharing common data typically provides the strong incentives and self-correcting feedback necessary to maintain high-quality data. Though data sharing usually raises some complex political and security issues, the benefits generally far outweigh the difficulties.

Data Independence

An important goal of database management is to increase the degree of **data independence** of the system—the separation between the database and the application programs. If, for example, a change is made in a data element, modifications should be limited to those application programs whose logic is affected by the changed element. A change only in the *physical* organization of the database, such as moving some records to a new disk storage device, should require absolutely no change in programs; instead, the DBMS should automatically translate a program's reference to a record into its current physical location. Data independence should work the other way as well: the system should minimize changes in the database necessitated by program modifications.

Data independence increases the flexibility of an MIS. In more conventional designs, where strong ties exist between data and programs, a change in the database may force major program modifications. The seemingly rather simple move from a five-digit postal ZIP code to a nine-digit code, for example, has caused major problems for many organizations. With data independence, changes would not be necessary except for those few programs whose logic is affected by the added four digits (although collecting the extra digits could be quite expensive, of course). In a world with changing technology and information needs, the flexibility achieved through data independence offers the only hope of keeping the system responsive.

Data Integrity, Security, and Backup

An organization's database may be the most valuable part of its MIS (other than the accumulated skills of the persons who build and run the MIS). The database is almost certainly likely to have greater value than all the hardware combined, and it may well be more valuable than the software. Clearly, the database merits very careful protection.

Protection has three aspects: **integrity, security**, and **backup**. Integrity is achieved by controlling the quality of the data that enter the database, security by guarding against unauthorized or inadvertent access to the database, and backup by providing procedures for reconstructing the database if necessary to recover from a major loss of data.

Integrity A comprehensive DBMS must provide mechanisms for protecting the integrity of entered data. The system needs to maintain information about the allowed set of values for a given data element (ideally as part of the data dictionary). This information is then made available to the data editing program that checks incoming data. An interactive data entry system can immediately flag any detected discrepancy so that the terminal operator can correct the error on the spot. In a batch system, the discrepancy can be identified automatically, but correction of the error must wait until the next processing cycle.

Security The database may contain highly confidential information having great strategic or commercial value. Sensitive personal data such as salaries or medical histories raise special problems of security and may even entail legal sanctions against unauthorized disclosure. The DBMS, along with the computer's operating system, must therefore provide a degree of data security that corresponds to the highly sensitive nature of the database. The difficultly of providing data security is exacerbated by the growing trend toward shared databases and distributed, interconnected systems in which potential threats can come from a myriad of remote terminals located at sites where protection against unauthorized physical access cannot be assured.

The operating system provides some security. When a user signs on to a terminal, the operating system usually requires some form of user identification, such as a unique user identifier for accounting purposes and a confidential password known (supposedly) only to the user.

The DBMS provides another ring of security. Consider the following mechanism, for example, which is a representative, though simplified, version of the security provisions built into a DBMS. The data dictionary associates a security code with each record type. The system also maintains a table with an entry for each authorized user that gives the security codes of the data to which a user is allowed access. When a user seeks access to a given record, the system checks to see whether the user has the authority to do so. If he does, the information is provided; otherwise, the request is denied. Other actions might also be taken, such as notification of security personnel if the system detects repeated attempts at unauthorized penetration.

The above description ignores an important detail: it speaks of "access" to the database without specifying the type of access. The system must be capable of differentiating between two types, access to *read* and access to *write* (i.e., change). Management might be quite willing, for example, to authorize an employee to read her own employee file, including her current salary; she should probably not have the right to change her own salary or to read the salary of any other employee. A user's security code must therefore specify whether it grants only read access, or write privileges as well.

This discussion omits many details about the exceedingly

complex topic of data security. An important issue, for example, is the granularity of the access barriers erected around portions of the database. Conceivably, separate access authority could be granted for each individual data element—a given employee's salary, say. This would be cumbersome to administer and inefficient to process, however. Access is generally authorized at a much more aggregate level, such as a specified collection of related data elements of a given type (e.g., all salary-related data pertaining to a given division of the company).

A secure system should allow after-the-fact auditing in addition to, or even in place of, before-the-fact access barriers. For example, the system can create an audit record each time a change is made to a data element. The audit record could include the identification of the data element and its values before and after the change. In addition, the audit record could identify the source of the change—the person making the change (known from the user's I.D. given at sign-on time), the location from which the change came, and a "time-stamp" that gives the time and date of the change.

When retained in a form that permits ad hoc analyses (e.g., by using the storage and retrieval facilities of the DBMS itself), the audit records permit a complete reconstruction of the events leading to the database's current status. A report could be generated, for instance, to show all the changes made to a given data element (a suspect account balance, say) by a given employee over a specified range of dates. The ability of the system to reconstruct events may permit the granting of relatively liberal write access from a variety of decentralized sources, with the assurance that errors or deliberate violations can be detected and corrected after the fact (assuming that in the meantime the perpetrator has not fled with massive ill-gotten gains).

Backup Finally, we consider the backup issue. Protecting the database requires a means to recover from a major loss of data due to (1) destruction or failure of the storage medium, (2) inadvertent or malicious introduction of bad data, or (3) improper deletion of data. This is traditionally done by taking a periodic "snapshot" of the database and storing the copy in a safe location (generally away from the data center). The system also retains all subsequent transactions that change the database. A destroyed portion of the database can thus be reconstructed by restoring the latest copy and reprocessing the subsequent transactions.

This works perfectly well with batch processing in which a copy of a file is created as a by-product of the transaction processing. The more contemporary environment with a massive on-line database presents considerably greater difficulties. One cannot shut down the world to make a copy of the database. Creating a backup has to be done on the fly while the system continues to operate.

Restoration of a lost portion of the database is much more difficult with an on-line system. A large system might process dozens or even hundreds of transactions per second coming from a far-flung distributed network. At any instant in time the system could be handling a great number of transactions in various states of completion. Some of these transactions may have already updated the database, while others have not. Recovering from a systems failure and restoring the database require keeping track of the current status of each transaction to avoid losing the update from a transaction or counting it twice by reprocessing a transaction that has already updated the database.

Backup protection exacts a price in terms of redundant resources and processing overhead. One approach, for example, is to continuously maintain two on-line copies of the database. By paying for 100% database redundancy, the system can instantaneously switch to the second copy if anything happens to the first.

In dealing with all aspects of database protection, designers face the inevitable tradeoff between value and cost: what is the value of added protection and how much will it cost? Advances in hardware and system software make it easier and cheaper to obtain a given degree of protection. As costs come down and an organization's dependency on the MIS grows, the tradeoff point moves inexorably toward an increasingly higher level of protection.

These are enormously complex issues, well beyond a detailed discussion here. User organizations must rely heavily on hardware and system software vendors to build in capabilities for coping with the complexities of backup and recovery. A great deal of progress has already been made, and these capabilities are being extended as the market demands ever-greater reliability in an on-line environment.

The Effect of DBMS-Based Tools on Development Strategy

A DBMS forms the core of most 4GLs. The leading DBMS now combine a variety of programming tools designed to provide a productive development environment. It is hard to overemphasize the importance of these new tools. Their importance comes not merely from their enhanced productivity, as valuable as that is, but also from the way they affect our view of the entire development process.

The traditional process places the emphasis on individual applications. The organization's overall information needs are identified through some sort of needs analysis that eventually ends up with a set of desired applications. These applications then become the primary focus of the implementation process.

With this approach, the management of data resources has a decidedly secondary role. Only after applications have been defined do designers typically give much thought to the specifics of file organization and the sources of data to support the applications. Passing data from one application to another tends to be awkward and complex, and so a premium is placed on keeping each application relatively free of links with other applications. Such data sharing that occurs tends to be handled on a pair-wise basis, with little overall standardization.

Figure 7-3 shows the consequences of this form of data sharing. The number of pair-wise links among them grows roughly in proportion to the square of the number of applications—in this example, six applications yield 15 pairs. Each pair requires some effort on the part of the two development teams to agree on such matters as data definitions, data formats, and the timing and volume of the data flow. The substantial effort re-

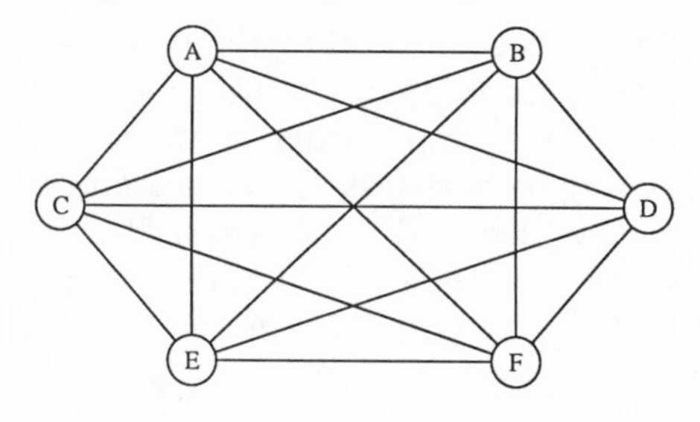

Figure 7-3. Pair-wise links among applicants.

quired to share data among applications discourages the attempt except in cases of the most obvious need (e.g., applications exchanging a large volume of time-critical data). Standard data definitions would simplify matters considerably, but would not materially reduce the fundamental barriers to data sharing intrinsic to an application-oriented design strategy.

Figure 7-4 shows the links among applications with the use of a shared database. Each application has a single interface with the database via the DBMS; consequently, the number of links that must be managed grows only in proportion to the number of applications. Even though no direct links exist among

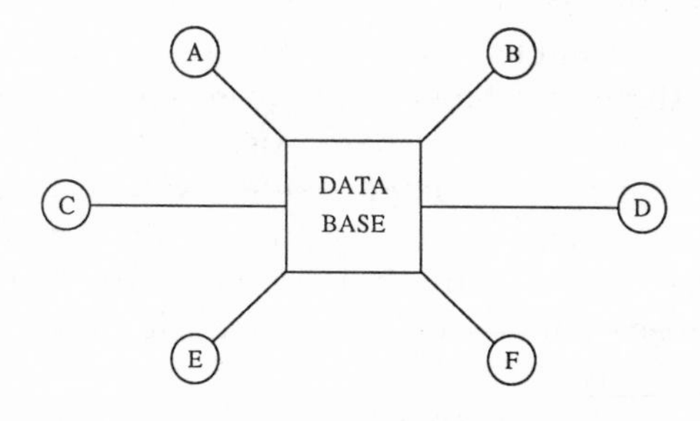

Figure 7-4. Links among applications through a DBMS.

applications, communication among them raises no problem as long as each application adheres to the standard data conventions. Program A, for example, can communicate with Program D merely by storing its output in the database in standard form; the output can then be read by Program D. Perhaps more to the point, the implementers of Program A can do their work knowing only about the outputs required by Program D; they need not know anything about the internal details of Program D.

Data sharing through a common database shifts the emphasis from applications to data. It is generally not necessary for designers to pay detailed attention to individual applications to establish the content and structure of the database; it is enough merely to define the general nature of the transactions to be processed by the system as a whole. An experienced designer, working closely with a sales manager, for example, would normally be quite capable of defining the data elements in a customer record without knowing all the applications that will eventually use customer data.

A focus on data resources in designing the MIS requires the preparation of a comprehensive description of the database, as discussed earlier in connection with the data dictionary. This is not a simple task, of course, but it is made easier by dealing with the database as an integrated whole rather than as multiple fragments associated with separate application programs. The amount of effort to define the database is largely independent of the details of transaction processing.

What does this mean in terms of implementation strategy? It implies that the implementation should concentrate on defining the database and building the core applications that feed the database. Once the database exists with high-quality data, new applications can be implemented with relative ease. The need for new data will always arise in the course of implementing new applications, but the flexibility inherent in the DBMS approach can generally accommodate these changes with no great difficulty.

Consider how this approach might be applied to our distributor of industrial products. The core processing functions for this business deal with entering a customer's order, preparing the shipment of available items, shipping the items, billing the customers, and managing inventory. Without knowing the details about each application, a competent designer can fairly readily identify the data necessary to describe customers, sales orders, inventory items, purchase orders, suppliers, and employees. Later on, as individual applications get implemented, the designers can concern themselves with the precise logic of the programs.

The sequence in which the applications are implemented depends on their relative payoff and the availability of input data. The order entry function could be an attractive beginning application, for example, because it provides useful services and generates the input data used in subsequent processing.

As discussed earlier, it is desirable to keep each application to the minimum feasible size. A process of serial implementation of relatively small chunks of the system offers the great advantage of reducing the development time and complexity of each increment. It also provides more immediate benefits from the early applications, and allows subsequent pieces to take advantage of the learning that occurs during the course of developing and using the earlier ones. Conventional wisdom suggests that the development time for an application should be limited to about six months.

Limiting each piece of a system to such small chunks is not always possible, especially for the initial parts of a system. In a tightly integrated system, a core of several applications may have to be installed together because of their mutual dependencies. In the case of on-line order entry, for example, if the system must check the inventory level of a requested item, the order entry function cannot be installed before the inventory control system. The inventory control system could possibly be developed first by itself if the existing order entry system can supply the required input data; otherwise, the two applications would have to be installed together as part of the initial core system.

The application programs that handle mainstream operational functions typically provide the bulk of the data used by more peripheral applications such as cost accounting. In this case, programming the accounting application rarely raises any great technical difficulties (although it might well be the occasion for resolving some tough management and accounting issues). Developers of the early core applications need pay little heed to the subsequent peripheral applications, and the subsequent applications have little or no impact on the earlier ones. This adaptive approach, made possible by current software development technology, is a key ingredient in a successful MIS.

Prototype Systems

A **prototype** system is an interim system used in the process of developing a more permanent version. At various stages of its development, or in different types of applications, a prototype system may offer capabilities ranging from a mere facade to an almost fully functional system that lacks only polishing and fine tuning.

Why Use a Prototype?

Like a mockup of a new product, a prototype system allows designers and users to examine, and often actually use, alternate designs. The prototype approach is based on the unassailable principle that none of us can specify long-term information needs in the abstract; we can only make an informed judgment about such matters when we see a concrete representation of the design. Anyone who has first-hand experience in designing a new house knows the value of a prototype: after living in it for a while, we invariably wish that we could have a second try to get it right (and then a third . . .).

The prototyping approach stands in considerable contrast to the conventional implementation process, in which the emphasis is placed on getting the specification right the first time. This "final" version is then implemented with a labor-intensive and inflexible technology. It matters little that this approach never works very well; all parties involved in the process—users and technicians alike—exchange recriminations and vow to do better the next time. In the meantime, the system undergoes a series of "maintenance" modifications that gradually converge on a more satisfactory design.

The prototype approach recognizes up front that changes inevitably occur. It thus places a premium on making the changes as easy, quick, and inexpensive as possible. Not surprisingly, prototyping has become practical and popular only with the availability of high-productivity software tools.

The Prototyping Process

In perhaps its most representative form, a prototype evolves through many versions until it eventually becomes a working application. In the early stages, the focus is usually on developing effective human interfaces and skeletal algorithms to deal with the central functions of the application. Typically ignored at this point are such things as machine efficiency, handling rare exceptions, and providing "bulletproofing" to make the system as invulnerable as possible to erroneous inputs. Although a successful production version must eventually cope with these matters, such refinements add greatly to the cost of development (well over half the cost in many cases) without contributing a great deal of insight about the primary functions of the application. Early versions get changed substantially over the course of an application's evolutionary development, and so one wants to avoid carrying all the baggage of a finely tuned program through the entire process.

It goes without saying that users must play a large role in the prototyping process. A prototype provides a concrete representation of the application infinitely more understandable to a user than reams of abstract documents. With a prototype, users have the opportunity to get the feel of a working system, understand its strengths and weaknesses, and request improvements and extensions. This "try before you buy" approach substantially increases the odds that users understand what they have agreed to and must eventually make work.

With each evolutionary step the prototype becomes more productionlike. When it settles down to a relatively stable and fully operational program, management must decide whether to continue to operate the program as a completed application. Alternatively, the then-stable design could be transliterated into a third-generation language that offers the advantage of greater machine efficiency or compatibility with the rest of the system. As 4GLs become more widely used in production applications, most prototypes are likely to evolve smoothly into a regular production version without the burden of having to be translated first into another language.

The full prototyping process does not apply to critical high-volume applications for which reliability, machine efficiency, and complex processing requirements are important design criteria. The core of an on-line system, for example, would generally not be susceptible to this treatment. Even so, certain parts of the system can usually be prototyped to good advantage. Reports, screen formats, and interactive user dialogues, for example, can be generated in mockup form using fictitious data so that users can examine proposed inputs and outputs of the system. Complex algorithms can also be worked out in a 4GL before committing them to a procedural language. Knowledge gained from the prototypes can be incorporated into the final design specifications with increased confidence that the production version will meet user needs. Following the prototype experiments, implementation generally adheres to the conventional life-cycle methodology.

Related applications outside of the high-volume mainstream can take full advantage of prototyping. This applies, for example, to programs that prepare periodic reports or handle relatively low-volume applications based on data generated by mainstream programs.

Problems with Prototyping

With all its advantages, the prototyping approach has its problems and hazards. In a culture that rewards careful needs analysis and stable specifications, not every MIS staff member finds prototyping a congenial problem-solving technique. Many find it difficult to cope with the ambiguity inherent in an approach that seemingly calls for experimentation and serendipity rather than thoughtful design and discipline.

Users, too, sometimes have problems with prototypes. On the one hand, they may only grudgingly tolerate the inevitable limitations of early prototypes; on the other hand, they may not want to bear the cost and delay of converting a fragile but apparently adequate prototype into a robust, reliable, and fully functional system. Furthermore, in an environment of continual change, users may eventually lose patience with a system that never seems to get finished.

These are not problems intrinsic to the prototype approach, but rather to a misunderstanding of its value and purpose. Use of a prototype should not be viewed as an excuse to implement half-baked ideas; even a prototype requires a careful—if not compulsive—needs analysis. Users must be educated to understand the role played by a prototype so that they will not be disappointed by its limitations or lured into a false sense of security by its strengths.

End-User Computing

Keeping up with the growing requirements for information is too heavy a load for the central MIS staff to carry on its own. The nontechnical user community must assume a larger share of the burden of satisfying its own needs. As in the telephone system, where "automatic" dialing assigned to users the task of direct entry of telephone numbers, user-friendly languages now allow users to do for themselves what formerly required the skilled hand of a professional programmer.

Advantages of End-User Computing

End-user computing offers several important advantages:

1. Users take over much of the programming effort from the overloaded MIS staff, thus freeing technical experts to concentrate on problems for which they have a comparative advantage.
2. Users can respond to their own changing needs without having to express these needs to a technical intermediary who may be unfamiliar with their applications.
3. Users can better set their own priorities, without having to contend for resources with other organizations or with central MIS management.
4. Users employ their own resources in satisfying most of their information needs; consequently, they are in an excellent position, with the right set of incentives, to balance the value of the information against the cost of supplying it.

What Is End-User Computing?

The term **end-user computing** needs some clarification. It is not limited to the case in which the ultimate consumer of information actually sits before a screen and interacts directly with a computer. The essential issue is organizational proximity and control, not who programs or operates the computer. End-user computing exists if programming and operating the computer fall within the organizational boundaries of the end-user of the resulting information. Thus, the marketing vice president engages in end-user computing when she requests her staff assistant to use the computer to prepare a special report.

In practice, distinctions between end-user computing and traditional computing get fairly blurred. A user who develops a stand-alone spreadsheet program on his own workstation for his own personal use is quite clearly practicing end-user computing. The situation is much less clear, however, when an administrative assistant uses an interactive terminal to invoke a parameterized report generator created by the central MIS staff and executed on the central machine. Nevertheless, both ends of the spectrum of end-user computing manifest a fairly clear shift of responsibility from the central staff to the user community.

End-User Computing and 4GLs

Like the professional programmer, end users benefit greatly from the productivity achieved with fourth-generation tools. The tradeoffs are somewhat different, however: the professional favors a powerful language that emphasizes productivity and ease of use, while end users generally value ease of learning and retention.

The professional can afford to spend considerable time to learn a programming language if it provides powerful and productive development tools. In contrast, an end user whose primary interests and skills lie in a functional area such as financial analysis or marketing can usually afford to spend no more than a couple of days learning a computer language. He may find only occasional use of the language, and thus must be able to retain an adequate level of skill without continual practice. Furthermore, to motivate him to devote the necessary time for learning a new language, the language must provide a distinct advantage over the user's best alternative means of satisfying a need—for example, unaided intuition, "back-of-the-envelope" calculations, or a previously learned computer language that is less capable than the new one but still adequate for most computational tasks.

Limitations of End-User Programming

The language requirements for end-user programming place a fairly severe constraint on what most end users can be expected to do for themselves. These users are generally limited to such tasks as a simple spreadsheet program or query specification. A more difficult task—for example, use of the procedural programming features in a spreadsheet language or DBMS—would generally require the assistance of a technical support person. (It should be noted, however, that some end users acquire great skill in specialized areas of computing, rivaling or even exceeding the skill of the professional staff.)

While a program developed by an end user may serve its purpose quite well, it should not be considered for use by others without a great deal of caution and care. The end user typically prepares the program solely for his or her own use, and generally has little skill or incentive to give much attention to such "niceties" as human engineering, testing and validation, user documentation, and maintainability. A professional programmer has usually learned through bitter experience that such considerations are essential for any program on which the organization relies, but this is not necessarily intuitively obvious to the amateur.

Because of these weaknesses, a user-developed program should be reviewed and modified (or completely rewritten) by the professional MIS staff before it is made available for use by others in any application that carries a significant risk if an error or failure occurs. An accounting or analytical program on which important decisions are based, for example, should certainly be subject to such a process. Usually a user-developed program can safely be granted access to *retrieve* shared data (assuming that the user has authority to look at the data), but would generally not be permitted to *update* the database for fear that an erroneous program might corrupt the data.

Role of the Central Staff

End-user computing should not be confused with unfettered anarchy. The central MIS staff has a critical role to play to mitigate the risks of undesirable duplication of effort, incompatibilities, and amateur programs. In performing this role, the staff must walk a fine line between encouraging imaginative and effective use of the computer by end users and avoiding some of their more egregious errors. The means for achieving a reasonable balance include the following services:

User education and training.

Preparation of user manuals.

Application consulting.

Operate a corporate **information center.**

Hardware and software selection.

Vendor negotiations and bulk purchase arrangements.

Technical support for selected products.

Maintenance of hardware and software.

Assistance in connecting to the communications network and obtaining access to shared data.

Promotion of user groups and self-help mechanisms.

Development of mechanisms for sharing and supporting user-developed programs.

In short, the central MIS staff should provide a supportive and facilitating environment that allows users to solve their own problems consistent with the needs of the organization as a whole.

General Guidelines for Software Development

The software problem will not go away; success in "solving" it only stimulates further demand and adds to the bottleneck. It is safe to assume that demand for new software is essentially insatiable. The task facing the organization is to manage the problem in a way that best meets its most pressing needs.

General guidelines for such complex matters always run the considerable risk of oversimplification, but the following ones at least provide a reasonable beginning point in setting a software development policy:

- Apply a fully disciplined software engineering methodology to only a small fraction of the organization's mainline applications (e.g., those that have a high-volume of use, serve a wide constituency, or meet a critical need).
- Use a conventional procedural language only as a last resort (e.g., within the central core of a mainline program that has critical efficiency or response time requirements, or elsewhere to handle complex functions that are infeasible to program in a 4GL).
- For most applications, emphasize responsiveness and reduced development costs through the use of fourth-generation tools, prototyping, and end-user computing.
- Be willing to develop and use an application that is less than perfect, and then improve it through an adaptive process that takes advantage of organizational learning.

Further Readings

Boar, B. H., *Application Prototyping,* John Wiley & Sons, 1984. A detailed coverage of the prototyping approach to systems implementation.

Date, C. J., *Database: A Primer,* Addison-Wesley, 1983. An introductory discussion of database management by one of the leading authors on the topic.

Everest, Gordon C., *Database Management,* McGraw-Hill, 1986. A comprehensive discussion of the design and policy issues of database management, with special emphasis on preserving the security and integrity of the database.

Martin, James, *Application Development Without Programmers,* Prentice-Hall, 1982. A discussion of fourth-generation development tools, written by one of the leading authors in the MIS field, who is also one of the strongest advocates for the 4GL approach.

Martin, James, *Fourth-Generation Languages,* vol. 1, Prentice-Hall, 1985. Similar in orientation to the above book, with discussion of some of the leading 4GLs.

McFadden, Fred R. and Jeffrey A. Hoffer, *Data Base Management*, Benjamin Cummings, 1985. An introductory textbook on database management, with a rich variety of examples drawn from business organizations.

❖ 8 ❖

Economics of Information

The Value and Cost of Information

Traditional economic theory treats land, labor, and capital as the three fundamental economic resources. Increasingly, we now recognize information as a fourth critical resource. Given the growing strategic importance of information technology, management needs to have a clear view concerning the economics of information—the way information adds value to the enterprise, and the relationship between the value and cost of the information.

Information provides value through such means as lower production costs, increased revenues, and better decisions. It costs money in the form of salaries, hardware, purchased software, communication services, space, and supplies. The goal in implementing an information system is to choose a design that yields the best overall balance between value and cost.

Characteristics Governing Value and Cost

The design of an MIS ultimately comes down to a set of specifications that deal with such matters as the content and timeliness of information outputs. Each characteristic of a system affects the value and cost of its outputs. An improvement in any characteristic adds to the gross value of information, but in general it also adds to the system's cost. Designers should seek the combination of specifications that provides the best net return.

Among the most important characteristics of concern to the designer are the following:

- Availability of relevant outputs (e.g, printed reports or interactive screens that satisfy users' information needs).
- Selectivity of outputs, aimed at reducing the probability of displaying irrelevant outputs or failing to display relevant ones.
- The variety of transactions handled by the system.
- The degree of "intelligence" embedded within the system (e.g., for automatic handling of complex transactions, such as setting the premium rates on nonstandard insurance products).
- The **timeliness** of the database: the interval of time between the occurrence of an event and the time at which the database is updated to reflect the event.
- The **response time** of the system: the time interval between a request for service and its completion.
- **Accuracy**: the degree of agreement between displayed or stored data and the correct value (not to be confused with **precision**, which depends on the number of significant digits used to represent a measurement).
- **Generality**: the range of functions available within the system to meet varied needs.
- **Flexibility**: the relative ease with which the system can be modified or extended to meet new needs.
- **Reliability**: the probability that the system will operate satisfactorily during a period of intended availability.
- Security: protection against loss of, or unauthorized access to, resources of the system (either accidental or intentional).
- Backup: redundancy to provide protection against catastrophic loss of portions of the system.
- User friendliness: the degree to which the system reduces any unnecessary burdens placed on the user in learning to use the system or in performing a desired task.
- **Robustness**: the extent to which the system reduces the vulnerability of the system to errors on the part of the user.

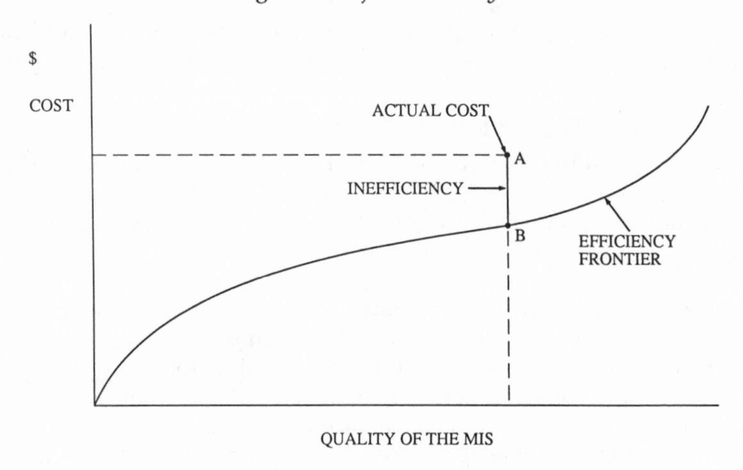

Figure 8-1. Cost as a function of quality.

Balancing Value and Cost

Designers must juggle the design characteristics of a system in reaching a compromise that best meets overall requirements. Their task is made more difficult by the fact that there are often **tradeoffs** between various characteristics. For example, the proliferation of interactive terminals for transaction processing reduces the update lag of an on-line database, but it also exacerbates the problems of security.

Technical means exist to achieve higher quality along any design dimension. Added reliability, for example, can come through such means as more extensive testing or by increased redundancy in hardware, communication channels, and power supplies. Reduced response time can come from additional processing or communications capacity, or by fine tuning the software to make it more machine efficient. Naturally, costs rise as the quality goes up—eventually very steeply as the limits of current technology are approached.

In Figure 8-1, the curve labeled **efficiency frontier** shows how costs rise as quality is improved. This needs to be explained. The cost of achieving a given level of quality—that is, meeting a given set of specifications—depends on the efficiency with which the system is implemented and operated. Suppose that System A and System B in Figure 8-1 have identical quality characteristics. System A, however, has a higher cost due to a combination of such factors as use of less efficient development tools, less efficient programs or storage organization, more expensive hardware or software vendors, less efficient operating procedures, and an obsolete design that does not take advantage of current technology.

The efficiency frontier shows the relation between quality and cost when using the most efficient available development and operating methodologies. This does not mean that any point on the efficiency frontier represents the theoretically most efficient system, because the design costs of discovering the theoretical optimum clearly have to be taken into account in measuring overall efficiency: beyond some point, the costs of fine tuning to improve efficiency exceed the savings in other cost components. Almost all systems offer some worthwhile opportunities for improvements in efficiency, particularly in view of the continuing possibilities presented by technological advances. The design and management effort required to make such improvements, however, can often be spent more productively by focusing on improvements in quality and not merely in efficiency.

Consider now the *value* side of the cost-value tradeoff. The gross value of the system increases as quality goes up. Eventually, however, incremental benefits begin to flatten out in observance of the economist's well-known law of declining marginal returns. Reducing a system's downtime by a factor of ten, from 2% to .2%, say, would typically add substantial value in an on-line order entry system; improving the system still further, by going to a .02% figure, for example, would generally result in considerably reduced (but still positive) incremental benefits.

The optimum point occurs where the net value is maximized—that is, where gross value exceeds costs by the maximum amount. (If value never exceeds cost, the whole issue becomes moot because the system should obviously not be built.) Equivalently, the optimum occurs where the marginal value just

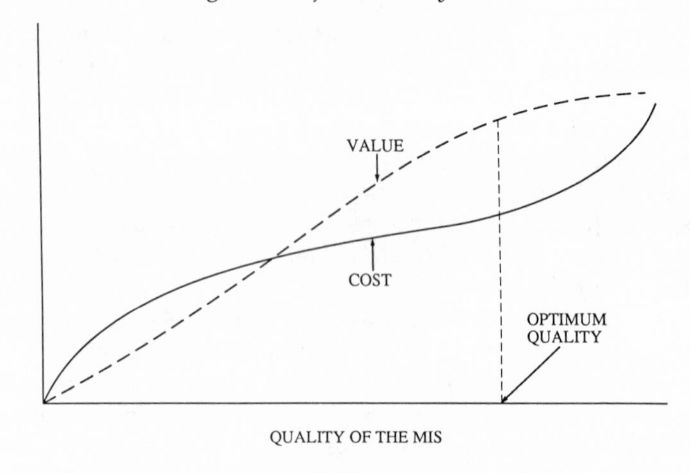

Figure 8-2. Tradeoff between value and cost.

equals marginal cost (under the stated assumptions of declining marginal benefits and accelerating costs), conforming to the common-sense notion that we should increase quality up to the point that it begins to cost us more than it's worth. Figure 8-2 shows a typical tradeoff between the value and cost of improving one of the design characteristics.

Figure 8-2 is related to some of the discussion in Chapter 6 about needs analysis. If you ask a manager to specify her information needs, it would be reasonable for her to respond by asking for any information that has a significant *gross* value. The objective is to maximize *net* value, however, which requires a consideration of cost as well as value—something the typical manager is not equipped to do very well. To strike the right balance between value and cost, it is necessary to elicit from managers some judgment about the shape of the value curve. That can rarely if ever be done in the form shown in Figure 8-2, but a workable approximation can be found by asking managers to categorize their information needs by relative value—for example, "must have," "should have," and "nice to have," as discussed earlier.

Efficiency vs. Effectiveness

Figure 8-2 illustrates an important concept—the difference between the **effectiveness** and the **efficiency** of a system. Effectiveness depends on the system's balance between value and cost, which is a principal management issue. Efficiency depends only on the cost of achieving a given set of specifications, which is a technical issue. This distinction is sometimes expressed in terms of "doing the right thing" (effectiveness) and "doing the thing right" (efficiency).

It is quite possible to implement an efficient system that does a poor job of dealing with the needs of the organization. A low-cost system that fails to provide valuable and well-justified decision making information, for example, could be efficient but not very effective. The same could be said of a "gold-plated" system that moves well beyond the optimum balance between value and cost.

It is difficult to argue against efficiency. After all, if the organization engages in an activity, it should try to minimize the resources devoted to performing the activity. Efficiency in information processing, however, is typically not of major strategic importance (unless you happen to be in the information processing business). Most MIS budgets fall within the range of 1 or 2% of total organizational costs, and so even if efficiency improvements eliminated all costs—not a likely prospect, to be sure—the net effect on the bottom line would generally be relatively minor.

An effective system, in contrast, can make substantial improvements in the mainline activities of the organization. The benefits can be highly leveraged and provide a significant strategic advantage. Management should thus concern itself primarily with issues of effectiveness, leaving to a competent technical staff the task of dealing with matters of efficiency.

Calculation of the Value of Information

The value of information comes from the incremental effect it has on the organization's performance. Suppose that we have a means of expressing performance—or *payoff*—in tangible monetary form, such as a reduction in cost or an increase in profit. The value of a given piece of information, such as a report, equals the increase in payoff that the organization derives from use of the information. That is,

Value of information = Payoff *with* the information − Payoff *without* the information

A Simple Example

Information value defined as the incremental effect on payoff is an important and practical concept. Unfortunately, however, it is difficult to calculate the value in practice because realistic problems quickly become very complex. Let us look at a very simple example to show what is required to perform such a calculation.

Suppose that it costs $1 million to launch a new product. If a product is successful it can be expected to generate future revenues having a net present value of, say, $12 million (including the effect of the initial $1 million cost); if the product fails, the entire $1 million is lost. Our past experience shows that we succeed with only 10% of our new products. If we were to develop ten new products, we would expect to succeed with only one of them and fail with the remaining nine. This would lead to a net gain of $3 million (one $12 million gain offset partially by nine $1 million losses), or an average gain of $.3 million for each new product. If we have no way of knowing in advance which products will succeed and which will fail, our best strategy is to develop as many new products as we can, reaping an expected gain of $.3 million for each such product.

Suppose, though, that we develop a market research technique that gives infallible forecasts of success or failure. When the market research study predicts a success (which will happen in 10% of the cases), we naturally choose to launch the new product and will thereby gain $12 million. When the study predicts a failure (in 90% of the cases) we will *not* launch the product and will end up with zero profit (and zero loss). What is the value of each forecast?

Out of ten products, one success and nine failures can be expected. This will lead to one $12 million gain and nine zero gains, or a total of $12 million for the ten products. On average, each product will yield a profit of one-tenth of this, or $1.2 million. The value of the forecast can therefore be calculated as follows:

Value of information = Profit *with* the forecast − Profit *without* the forecast
= $1.2 million − $ .3 million
= $ .9 million

Some General Conclusions About the Value of Information

Let us examine in more detail how information increased profit in the simple forecasting example. It can be broken down into three steps:

1. A prediction has some "surprise" content, in the sense that it changes the decision maker's view of a product's prospects; a prediction of failure, for example, changes the decision maker's *prior* probability for failure of .9 to a *posterior* probability of 1 (i.e., certainty).
2. A prediction sometimes causes a change in the decision to launch a product: a prediction of failure shifts the decision from "launch" (which would have been the best action in the absence of a forecast) to "no launch."
3. The changed decisions lead to a higher expected payoff—from $.3 million to $1.2 million.

The extent to which new information affects a decision—and hence has value—depends on the particular circumstances. A forecast of success has a high "surprise" content (since it shifts the estimated probability of success from a low value to certainty); nevertheless, it does not affect the launch decision, because the optimal decision is to launch in the absence of a forecast of failure. If a forecast predicts failure, however, it alters the decision from "launch" to "no launch." The information derives its value from these altered decisions (by avoiding a $1 million loss for 90% of the products).

In real life, of course, predictive information used in decision making is seldom perfect. If a forecast predicts a success, the product will sometimes fail; if the forecast predicts a failure, in some cases the product would have been successful if it had been launched. The expected value of each forecast can be computed if its accuracy is known, but we need not spend time here to go through the fairly laborious calculations. As you would expect, the value of information generally declines (and never increases) as accuracy is reduced.

Estimating Information Value in Practice

For real-world problems, we rarely have sufficient knowledge to place a hard monetary estimate on the value of information. To actually determine the value of market research information, for example, we would have to quantify each of the three steps by which information acquires value. Thus, we would need to know:

- How would the information change our view of the market—for example, in our estimates of the probability of achieving $50 million in sales for a newspaper advertising expenditure of $3 million?
- How would the changed estimates alter our decisions—would it cause us to shift advertising from newspapers to TV, for example?
- What would be the expected increase in net profit from the changed decisions?

Often the organization has only a hazy idea about such matters as its potential market size, price elasticities, and the response from various advertising media. Even less could it estimate the likely "surprise" content of additional market research information.

Even if a formal estimate of the surprise content could be obtained, the organization would still have to determine how the information would affect its advertising decisions. This would be tantamount to having a formal decision model that relates the proper level of advertising for each medium and location for each estimated level of potential demand. Such formalization is clearly not generally available.

But even this would not be enough. The final step in the evaluation process would require an estimate of the net profit from each alternative decision. This could be quite difficult, possibly requiring consideration of such detailed matters as capacity limitations and the marginal cost of production. In more complex situations with multiple goals—profit, share of market, image, rounding out the product line, and the like—translating the payoff into a monetary estimate can be virtually impossible.

Making judgments about the value of information is not hopeless, however, as long as management is not compulsive about getting hard monetary estimates. A number of approaches exist for performing cost-benefit analyses of information systems. Often it is possible to justify the development cost on fairly hard evidence. In other cases, though, this is not possible, and management must use largely subjective means in judging whether to install a given system.

In assessing a possible expenditure for information, either through quantitative evaluation or subjective judgment, it helps to break the issues down into the three value-adding steps. Suppose, for instance, that we are considering the value of on-line *updating* of the inventory file as part of a new order entry system. The alternative to on-line updating might be batch updating each night (most likely with on-line *retrieval* during the day). The problem can be analyzed as follows:

1. What additional surprises result from immediate (versus

24-hour) updating? If significant and unpredictable events are likely to occur over a 24-hour period (e.g., the sudden withdrawal of a large order from stock), then immediate file updating will provide significant surprises when such events occur. If, on the other hand, major events rarely occur without warning, immediate updating would not change very significantly the organization's view of its inventory position as provided from the previous day's updating.

2. What decisions will be altered if a surprise occurs? In some cases having current information may prevent accepting an order for immediate delivery after the stock has already been sold during the current day, or rejecting an order when a replenishment order has been received earlier than expected on the current day. It is quite possible, though, that rapid updating will have little effect on decisions. This could happen, for example, if tradeoffs favor setting inventory stocking policies so that stockouts rarely occur even in the face of unexpected withdrawals or receipts; under these circumstances, updated information would normally merely confirm that sufficient stock is available to satisfy an incoming order. Furthermore, on-line updating may not lead to many altered decisions if most customers are willing to place orders subject to backordering if stock is not immediately available.

3. What is the effect on payoff of an altered decision? If a surprise is sufficient to change a decision and avoid an error—rejecting a customer order when a stockout has occurred, or accepting an order when stock has unexpectedly been replenished—the payoff could be substantial. This would be true, for example, if an unnecessarily rejected order carries a high cost in lost profit, or if an unexpected delivery delay causes serious customer ill will (and hence the loss of future sales and profits). In many inventory systems, however, the payoff from avoiding an (occasional) error may be fairly modest.

Even though the value of on-line updating may not be too great in a typical inventory system, one should certainly not jump to the conclusion that rapid updating of the database is generally not justified. Consider an airline reservation system, for example, which deals with a particularly perishable product, a seat on a given leg of a flight. In this system, many surprises occur in a 24-hour period (requests for flights, canceled flights, mechanical or weather problems, etc.). Furthermore, the airlines attempt to operate with a high load factor, so any surprise might well alter an airline's decision (accept a request for a flight if a seat is available, reject it if not). The payoff from avoiding an error (revenue lost forever or a bumped passenger, say) can be quite high. Under these particular circumstances, on-line updating has an extremely high value that can easily justify the very high cost of operating an on-line reservation system.

In assessing the value of information, one must consider the combination of the three value-adding steps, not any one step by itself. In an order entry system with a high volume of incoming telephone orders, for example, the expected value of up-to-the-minute information may be low for each order, but the system may still yield an attractive return when the value per order is multiplied by a large number of orders. Similarly, a system for monitoring catastrophic failures in a chemical plant or nuclear power station may deal with exceedingly rare events, but its expected value may nevertheless be extremely high owing to an astronomical payoff if a rare event does occur (avoidance of injuries and deaths through the rapid initiation of emergency procedures, say).

A final point should be emphasized here. Even though in practice it is generally infeasible to quantify the value of information, subjective judgments can be substantially aided if decision makers think about the problem as a three-step process. A manager, when faced with the decision about the likely value of new information, should ask:

1. What surprises will I receive from the information, and how often will I receive them?
2. What decisions will I change for the better if I receive such a surprise, and how often is this likely to occur?
3. What will be the payoff from making these improved decisions?

Cost-Benefit Analysis of Information Systems

Implementing an information system requires an investment from which future benefits are derived. As in the case of other types of investments, management must determine the appropriate level of expenditures for information. In principle, if not in practice, this is a straightforward decision: spend up to the point that the marginal return from an expenditure for information just equals the marginal return from the best alternative use of resources.

Project Definition

One of the issues in performing this assessment is the granularity of the investment decisions. Investment opportunities in information systems are typically broken down in terms of **projects** having definable benefits and costs. The objective of cost-benefit analysis is to determine whether or not the expected benefits to be derived from a project are sufficient to justify its development and operating costs.

Defining the boundaries of a project is not all that easy. Ideally, designers should define a project so that it is relatively independent of other projects, thus enabling management to evaluate it on its own merits. In practice, though, a project is likely to have significant links with other projects, and so the project's definition is often somewhat arbitrary. A project to develop an order entry system, for example, would typically interact with projects to develop such applications as inventory control, production scheduling, and accounts receivable.

Nevertheless, the organization cannot deal with an entire management information system as a single monolithic lump; it must instead define (and install) the system on a project-by-project basis, with each project justified largely on the basis of its own costs and benefits. If a project cannot be justified on its own, but other attractive projects require it for their implementation (as a source of data inputs, say), then it can always be combined with the other projects and evaluated as a single composite project (even though the implementation of each subproject is managed as a separate task).

Project Size

Projects can vary greatly in size. The majority typically involve fairly minor "maintenance" changes that may require only a few man-days of effort. One organization found, for example, that half of its projects consumed less than 2% of its total development manpower. Projects of this sort certainly do not call for an elaborate cost-benefit analysis; resources should be assigned to them on the basis of a simple "back-of-an-envelope" analysis or an informed subjective decision. As end-user programming becomes more prevalent, these small tasks will shift largely to the user organizations, which are in the best position to make rational resource allocation decisions.

At the other end of the spectrum, a few large projects consume a highly disproportionate share of the development resources. These large projects require very careful cost-benefit analysis. The stakes can be very high, in terms of the cost of the selected projects as well as the opportunity cost of the scarce resources denied to other projects.

Estimating Costs and Benefits

To the technical staff falls the task of estimating the cost of a project with a given set of characteristics (specified outputs, response time, accuracy, etc.). This requires sketching out the design of the system, and developing an estimate of the resources necessary to implement the design. These resources may include such things as manpower (including the time of managers and users), hardware, purchased software or databases, consulting services, communication services, space, and supplies. Cost estimates must cover not only the development costs, but also the operating and maintenance costs over the full life cycle of the system.

Although cost estimation is usually a fairly straightforward technical task, it is nevertheless subject to some uncertainty due to unresolved design issues and errors in estimating resource requirements. At the earliest stage of a project, when feasibility is first assessed, cost estimates might easily err by 25%. As implementation proceeds, however, and the design becomes more refined, management should expect quite accurate cost estimates of development costs—a tolerance of 10% is usually a perfectly reasonable goal. Estimating the operating costs several years ahead may be subject to considerably greater error, but the cost-benefit performance of a project is often fairly insensitive to modest variation in costs that occur well into the future.

Estimating the benefits of an information system project generally presents a much tougher problem than estimating its costs. Putting a value on the system's outputs can be very difficult, as we have already seen. Dealing with benefits also raises some new organizational issues. Unlike cost estimation, the problem is no longer a technical exercise that can be handled largely by the technical staff. Most benefits come from effects on the organization outside of the information processing area, and so management and the user community must participate heavily in assessing benefits.

Analysis of a Cost-Reduction Project

Consider first the simplest case in which a project aims to reduce information processing costs by replacing existing equipment with lower-cost hardware, or by fine-tuning existing software. If only the efficiency of the system is affected, and not its effectiveness, no conceptual problems arise in assessing the increased value of the information outputs—by definition, the value remains unchanged. In this case, then, one must merely estimate the reduced costs over the life of the system and determine if these benefits justify the cost of the project.

This determination requires a classical investment analysis. A variety of criteria can be used, including payback period, internal rate of return, and net present value. For consistency in our discussion, as well as for sound theoretical reasons, we will stick to net present value as the appropriate measure of an investment's attractiveness.

With a time-phased profile of net cash flows over the life of a system, it is a simple matter to calculate the net present value of a project. Even though tax accounting may treat some (or most) of the costs as expenses rather than capital expenditures, a project should certainly be evaluated as an investment. If the net present value of the cash flows is positive, the project is worthwhile; if negative, it is not.

Analysis of Tangible Benefits

If benefits come from the increased value of a system, rather than merely reduced costs, matters become considerably more complex. One must get into the issue of the value of enhancements such as increased functionality, reduced response time, and greater flexibility of the system.

Under some circumstances it is possible to translate these improvements into a reasonable estimate of monetary benefits. One of the greatest potential benefits comes from cost reductions outside of the information processing system. Some examples might include the following:

- A reduction in inventory levels made possible by telecommunications links with field warehouses.
- Reduced capital costs achieved through more efficient scheduling of a hospital's operating room facilities.
- Reduced labor and fuel costs for a fleet of trucks by means of an optimizing program that minimizes the total distance travelled in delivering material from warehouses to customers.
- Reduced idle cash from tighter cash management.
- Reduced interest costs from the optimization of leasing terms.

A somewhat arbitrary, but important, distinction is made here between a cost reduction in information processing and one in another activity of the organization. From the perspective of the information system, the former benefit increases the system's efficiency while the latter increases its effectiveness. Although the *net* benefit is the same whether the improvement is viewed as a cost reduction or a value enhancement, the distinction can have practical significance.

Unlike an improvement in efficiency, an increase in effectiveness generally requires an increase in the information processing budget—presumably with offsetting benefits in other parts of the organization. If management pays undue attention to the data processing budget rather than net benefits, value-enhancing projects suffer a disadvantage in competition for resources with projects aimed at processing efficiency. This can have undesirable consequences, because most of the really significant improvements coming from information technology lie in greater effectiveness rather than increased efficiency.

Estimating the benefits that come from greater efficiency in operations may be analytically complex but this usually does not raise any great conceptual difficulties. Estimating the monetary value of reduced inventories, for example, may require considerable effort (such as determining storage and handling costs, obsolescence and wastage, and the cost of capital), but not much deep theory. The most serious problem is often estimating, in advance of implementing the system, the improvements that a new system will bring. However, analytical techniques exist for making such estimates—for example, analysis of a small test sample, extrapolating from the experience of others, use of a simulation model—which generally give results adequate for purposes of evaluating a proposed project.

Analysis of Intangible Benefits

The real difficulty comes in trying to assess an **intangible** benefit—one that cannot be translated directly into monetary terms. Examples of such benefits are improvements in product quality, improved availability of stock, reduced delivery times, enhanced customer services, greater plant safety, faster communications within the organization, and better management information. All of these are desirable benefits, and in some cases may be of crucial strategic importance, but expressing the benefits in creditable monetary terms may be impossible.

Take the case of improved availability of inventory. The improvement may very well result in greater sales, which, in turn, yield greater profits. The relationship between stock availability and customer demand is generally not well understood, however. We are thus forced to deal with the analysis in a less direct way.

Continuing with this same example, the following techniques could prove useful:

1. Quantify in nonmonetary terms. Even if improved stock availability cannot be translated into a *monetary* benefit, it is nevertheless possible in most cases to quantify some aspects of the benefits—the percentage change in stock availability, say, or the effect on average delivery times. This will at least provide a rational basis for exercising good subjective judgment, and will give a basis for measuring the actual results eventually achieved against the expected results.

2. Estimate monetary benefits from associated effects. A benefit that defies attempts to translate it completely into a monetary saving may nevertheless have associated benefits to which a tangible value can be assigned. An increase in stock availability, for example, might reduce the cost of expedited shipments made at premium freight rates. Although this tangible benefit may be relatively minor compared to the intangible benefit of improved customer service, it should certainly not be ignored in the analysis.

3. Determine boundary ("worst case/best case") estimates. Even though it may not be possible to get an objective estimate of a benefit, it is sometimes possible for a manager to give an estimate of an upper or lower bound. This may be all that is necessary to resolve the issue of whether a project should be pursued. For example, if the lower-bound (worst case) estimate of the value of improved stock availability is $500,000 per year,

and this is a sufficient amount to justify going ahead with the project, then a more refined estimate is not necessary. Alternatively, if the best case estimate were $100,000, and a value of at least $250,000 were needed to justify the project, it could be rejected without further analysis.

4. Express the cost in break-even terms. Suppose that justification for the project required an attributed value of at least $250,000 for the improvement in stock availability. This figure could be presented to management, which might then be in a good position to judge whether the likely benefits exceeded the break-even level. This might be more meaningful to management if the break-even point were expressed, say, in terms of the increase in sales revenue necessary to justify the project (e.g., a 1.2% increase in sales would pay for the project).

5. Tradeoff with a tangible benefit. Suppose that management has the option of keeping the stock availability at its existing level and taking the benefit from an improved inventory system in the form of a tangible saving—for example, a reduction in annual inventory carrying cost of $400,000. If management elects improved service over lower costs, the value of the intangible benefit can be set (at least) at $400,000, since this is the tangible benefit that management is willing to forgo to gain improved service.

6. Use the cost of the lowest-cost alternative. Suppose that management felt that it had to improve stock availability to respond to market pressures. In the absence of the proposed new inventory system, the lowest-cost means of improving stock availability might be to increase the average size of the inventory at an added expense of, say, $1 million per year in carrying costs. In estimating the benefits of the new system, then, a value of $1 million per year should be attributed to the improved stock availability (*provided*, of course, that failure to implement the new system would force management to carry the larger inventory).

Analysis of Composite Projects

Most projects provide a mixture of tangible and intangible benefits. To the extent possible, a monetary value should be placed on the benefits, including those incidental to improvements that are basically intangible in nature. If the project can be justified on the basis of the tangible benefits alone, then no further attention need be paid to the intangible effects (although they should still be included in the project documentation for eventual comparison with the actual results achieved). If, on the other hand, the project cannot be justified solely on a tangible basis, then the *net* tangible costs—the difference between tangible costs and tangible benefits—must be weighed against intangible benefits (using the techniques discussed in the previous section).

Suppose that the inventory control project discussed above has the cash flow profile shown in Figure 8-3. Implementation takes 18 months and costs $1.2 million. Peak development activity occurs during months 4 through 9, at a rate of expenditure of $96,000 per month. The core system begins to provide useful services at the beginning month 19, and takes six months before it becomes fully operational. The operational life is estimated to be about six years, with an assumed gradual decline in value during the sixth year. At its peak value, during months 24 through 78, the system generates a net tangible value per month of $10,000 (primarily from reduced inventory carrying costs, say). If a discount rate of 1% per month is used, the net present value of the tangible costs and benefits for this project is minus $721,000—clearly not a good investment on tangible grounds alone.

What value would have to be attributed to improved customer service for this project to become attractive? An intangible value during its full operational life of about $19,000 per month yields a zero net present value for the project. This is equivalent to saying that the project is worthwhile if the intangible benefit of improved stock availability is judged by management to have a value of at least $19,000 per month. Looking at the decision in another way, if the average profit contribu-

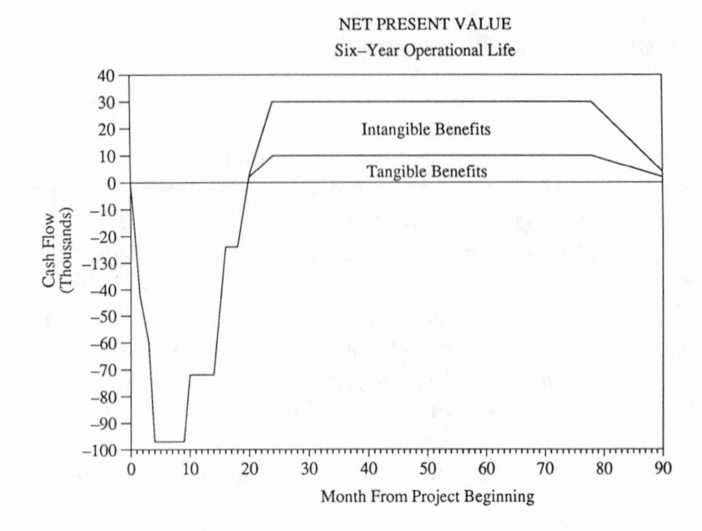

Figure 8-3. Cash flow profile of a project.

tion on sales is, say, 10%, an increase in sales of more than $190,000 per month would justify the project.

Sensitivity Analysis

All the estimates used in this analysis are subject to some error. Any cost-benefit study should include a sensitivity analysis that shows the effect of such errors.

For example, management might be interested in the break-even level of intangible benefit if the development time stretched out two additional months at the peak staffing level, adding almost $200,000 to the development cost. (It would be about $24,000 per month, an increase of 26% over the base estimate). Similarly, management might want to know the effect on the breakeven value of a seven-year life rather than six years (about $15,600 per month, a decrease of 17% over the base). Sensitivity information of this sort provides useful insights into the vulnerability of an implementation decision to errors in estimating project parameters.

Setting Priorities Among Competing Projects

Few organizations have the capability to simultaneously pursue all projects that pass the net present value hurdle. Several constraints limit the rate at which systems can be implemented:

- Sufficient technical staff personnel are generally not available to implement all worthwhile projects, or even to manage the implementation by others (e.g., outside consultants).
- Management cannot, or will not, increase the size of the technical staff—sometimes for budgetary reasons, but often because it does not wish to hire, train, and manage a larger staff.
- The most fundamental limit of all, perhaps, is the rate at which the organization can absorb the learning and change associated with new information systems.

Faced with more worthy projects than it can implement, the organization must set priorities for projects. One way of doing this would be simply to rank the projects by net present value. There are a number of objections to this practice, however.

1. The cost and benefit values used in the NPV calculations do not embody all the relevant information about a project—political issues and uncertainties, for example, may be difficult or impossible to incorporate into the analysis.
2. Interdependencies among projects may affect the cost of implementation, and hence should be taken into account in choosing the sequence of implementation.
3. All of the many resources needed to implement each project—people of various skills, specialized hardware, and the like—are represented in the NPV analysis by aggregate cost figures; in fact, though, some projects may require a disproportionate

share of scarce resources (e.g., telecommunication specialists), rendering such projects relatively less attractive (or even infeasible) compared to other projects with a lower NPV.

4. A simplistic ranking of projects by NPV does not take into account scale effects—for example, a large project may have a larger *absolute* NPV than a smaller one, but the *relative* return might be much greater for the smaller project.

Even with these objections, an NPV analysis has a useful role in setting priorities. One way of adding to its value is to consider explicitly the NPV per unit of constraining resource for each competing project. Suppose, for example, that we wish to rank two projects, A and B, and that the most limited resource is system design manpower. The projects have the following characteristics:

Project	*NPV*	*Design Man-Months*	*NPV per Man-Month*
A	$200,000	8	$25,000
B	$600,000	30	$20,000

Even though project B has the higher NPV, it should have a lower rank (all other things being equal) than project A because the latter yields a higher return per unit of the scarce design resource. This procedure gives, in general, only an approximation of the theoretically optimal solution, but it conveys the general idea of optimization subject to a resource constraint. To choose among several projects subject to a variety of limiting resources (designers, telecommunication specialists, computer capacity, etc.), one could use a mathematical procedure called integer programming (a variant of linear programming). It is seldom employed in project selection, however, because of its added complexity and the fact that the theoretical "optimum" solution is almost always subject to considerable management modifications to take account of factors not incorporated in optimizing model.

Despite its limitations, formal quantitative analysis of costs and benefits provides a useful input into the priority-setting process. However, it does not relieve management of responsibility for establishing the general strategic direction of systems development. The more strategic applications tend to deal with such issues as the long-term competitive position of the firm, basic product or service innovations, and management effectiveness. These are generally not matters subject to much quantification—certainly not in monetary terms. A policy that insists on investing only in projects yielding a demontrable tangible return is likely to foreclose important strategic opportunities. More will be said about this in the final chapter.

Relation of Cost-Benefit Analysis to Project Management

Cost-benefit analysis should not be viewed merely as a one-shot process to decide whether or not a project will be implemented. Rather, it should be incorporated as an integral part of managing an application's entire life cycle.

Cost-benefit analysis must cope with an inherent paradox. One needs to estimate a project's life-cycle costs and benefits to decide whether to proceed with the implementation. To get reliable estimates, however, one must proceed with the implementation at least through the gross design stage. Management of the implementation process must recognize the inherent uncertainty of the early cost and benefit estimates.

The practice of **creeping commitment** takes advantage of information that accumulates during implementation. Each stage has a well-defined budget, schedule, and set of deliverable outputs. Included in the outputs are refined estimates of the full-cycle costs and benefits. At the end of each stage, before proceeding to the next one, an explicit decision is made to continue with the implementation as planned, modify the plan in light of the latest information, or abandon the project altogether. A commitment to continue should extend only through the next stage, about which little uncertainty should then exist regarding costs, schedules, and deliverable outputs.

Major modifications or abandonments should come early in the process, when mistakes are relatively cheap. A variety of alternatives should be examined during the feasibility study and the early analysis and design stages. Since each alternative examined entails a certain amount of effort, the number must be limited to three or four substantially different alternatives. It is important, though, that they encompass a wide range of characteristics so that attractive options will not be overlooked.

At the low end of the range, a bare minimum design should aim only at satisfying the "must have" user needs (with perhaps some of the more cost-effective "should haves" thrown in for good measure). A high-end system should meet most of the "should have" and many of the "nice to have" needs. This wide range is quite likely to bracket the design that provides the best balance between costs and benefits.

As the implementation proceeds, the range of alternatives narrows quickly. Alternatives that lose their attractiveness on closer inspection should be abandoned early in the process. Effort can then be concentrated, at a much more detailed level, on the few surviving designs. Eventually the winning design emerges from detailed studies that consider the cost-benefit tradeoffs among system characteristics.

Cost-benefit analysis of a project should not cease with its deployment. Each of the inevitable modifications to the system should be viewed as a project subject to an assessment of its costs and benefits. In addition, it is desirable—though often quite difficult because of numerous confounding changes in the environment—to measure a system's actual costs and benefits compared to its planned performance. This not only motivates good practice and unbiased estimates, but it also provides valuable feedback information for improving the estimating process.

Charging for Computing Services

Difficulties in Determining Costs

We saw earlier that the analysis of costs generally presents relatively few conceptual difficulties. An exception to this, however, is the allocation of costs among various applications and users. Cost allocations can have an important effect on the way users view the tradeoffs between costs and benefits.

Strong arguments exist for charging users for the computing services they receive. A user receiving "free" services tends to overconsume them, with the result that costs to the organization exceed the benefits gained. Data center management, too, has no great incentive to strive for efficiency if users do not pay for information services and consequently do not exert pressure for lower costs.

In determining their demands for information services, users should aim to match costs with benefits. The incentives for doing this are generally most effective if the recipient of the benefits also bears the costs. A rational **chargeback** scheme of this sort fosters efficiency and effectiveness on both the demand and supply side—that is, on the part of both users and data center management.

Determining user charges raises a host of complex issues. The simplest case occurs when a dedicated system serves a single user, or a group of users within a single cost center. The problem of costing becomes much more difficult when users from different organizational units share common hardware and software, creating the need to allocate costs among these multiple users and applications.

One of the difficulties in such allocations is the treatment of "fixed" costs. An important characteristic of sophisticated information systems is that they tend to have a high **fixed** cost and a low **variable** cost per unit of capacity. Although capacity—along with its associated costs—can always be adjusted over a sufficiently long period of time, costs remain largely fixed once the capacity is established, regardless of the intensity of use. Up to the effective limit of capacity, then, the cost per unit of output is inversely proportional to the volume of usage. The cost of a given application thus depends on the total utilization of the system, not just on the application itself.

If only one job at a time runs on a computer, each job can be allocated all costs for the duration of the time it is on the machine. In a shared system with many simultaneous users, however, different jobs compete for different resources at the same time. The jobs tend to require a different mix of resources—an engineering calculation may require considerable CPU time, for example, while a typical data processing task spends much of its time on input-output operations.

How does one determine the cost of a job run in this kind of shared environment? This is like asking how to calculate the cost of moving a bushel of wheat by rail from Kansas to Los Angeles. In both cases the answer is the same: with difficulty. This classic problem of cost allocation has no completely satisfactory solution.

The conventional means of charging for a job run on a large mainframe computer consists of the following steps:

1. All the various costs of running the computer center are estimated for the time period in question—the forthcoming budget year, say, if prices are set in advance on an annual basis.
2. A set of resources is selected to serve as the basis for charging for work; these typically include CPU time, input-output volume (e.g., the amount of data transferred between main memory and disk or tape storage), the volume of data stored on disk, and the number of lines printed.
3. Estimated costs over the time period are allocated to each of the resources, and generally include allocations of such indirect costs as data center management, computer operators, space, and power.
4. The allocated cost of each resource is divided by its estimated usage to determine its charging rate; for example, if the total cost attributed to the CPU is $100,000 over the next year, and 4 million seconds of productive CPU usage are estimated for this period, the charging rate for the CPU is $100,000/4M, or 2.5 cents per second.
5. The usage of each resource is measured when a job is run; this

usage, multiplied by the appropriate charging rate and summed over all resources, then determines the cost of the job (e.g., the CPU component of a job requiring 100 CPU seconds at 2.5 cents each would be $2.50).
6. Users are presented with periodic bills for all jobs run during the billing period; these bills typically include a breakdown of costs itemized by resources consumed.
7. The difference between the total amount billed to users and the actual cost of operating the computer center constitutes the **variance** for the center.
8. The variance is disposed of by (1) absorbing it as an overhead cost for the period, (2) rolling it forward as an adjustment for the charging rates during the next period, or (3) retroactively changing rates and adjusting user bills so that the computer center ends up with a zero variance.

This chargeback scheme suffers from a number of serious defects:

- The system is complex, expensive to administer, and difficult for users to comprehend.
- Users generally have considerable difficulty relating their use of the computer to the itemized charges they receive.
- The costs levied against a given user depend not only on his own use of the system, but also on other factors over which he has no control (e.g., the efficiency of the data center and the total load on the system); consequently, the user may find it difficult to make rational tradeoffs between the cost and value of the services received.
- Rates are set at a level inversely related to expected demand, leading to undesirable user behavior—for example, if demand goes down, the rates for the remaining users go up, thus further depressing demand.
- Retroactive adjustment of rates to distribute the residual variance has several perverse effects: (1) it patently serves no useful managerial purpose in allocating resources, because users have no control over their past usage; (2) it makes planning and budgeting very difficult for users, since they do not know the charges until after their consumption of services; and (3) it does not encourage efficiency on the part of the data center, because the data center passes on to users all the effects of its own inefficiencies.

Even the concept of chargeback, let alone the particular mechanisms used, can be flawed under some circumstances. Although the goals of effectiveness and efficiency are generally best served by charging users for information services, some harmful effects can stem from a rigid demand to charge out *everything* in the MIS budget. Consider the case, for example, of a user department serving as the host site for a pilot application that has potential value throughout the entire enterprise.

If the application involves new and untried technology or operating procedures, considerable learning costs may be incurred. Such a project may also entail a significant risk of disappointment (e.g., higher costs or lower benefits than expected). The incremental learning costs and uncertain payoff may render the application unappealing to the individual host user, even if the likely benefits for the organization as a whole make the new application very attractive. A user therefore has no incentive to serve as a guinea pig, bearing all the costs and risks of pioneering. A strict chargeback policy thus can substantially reduce the organization's ability to engage in new or risky applications that could potentially have great strategic value.

Guidelines for Setting Chargeback Policies

The way in which the organization allocates costs for information services can have a major effect on its ability to develop effective, efficient, and imaginative systems. Chargeback policies must be tailored to the needs of each organization, but the following general guidelines raise many of the important issues:

- Each user should normally bear the costs of the information services provided.
- Costs charged to users should generally be based on long-term **marginal costs**—that is, they should exclude costs that will not change significantly over any plausible range of activity, such as the cost of the core management team for the data center.

- A small portion of the total central MIS budget—in the range of perhaps 5 to 10%—should be made available for innovative and possibly risky projects that no user department is willing to fund completely on its own.
- Revenue received from users by the data center should become discretionary "revolving" funds available to support the operation without further budgetary approval.
- Costs not controlled through user charges—that is, the fixed cost component of the data center budget plus centrally-funded exploratory projects—should be controlled directly at the corporate level.
- **Standard** rates, rather than "actual" (retrospective) costs, should be used as the basis for charging users; these rates should be based on expected volumes and an acceptable level of efficiency over the forthcoming accounting period (a year, say).
- Any variance at the end of the accounting period should normally be absorbed at the data center level, with due management attention paid to the source of the variance.
- Variances stemming from multiperiod effects (e.g., start-up of a new computer having excess initial capacity) can be carried over to the following period in setting the new rates.
- Users should be charged for any resource dedicated to their specific use (such as a disk drive devoted solely to a nonshared database); multiyear contracts with users of such resources often make particularly good sense.
- **Unbundle** charges—providing a separate "a la carte" price for each individual resource—so that users pay for the services they actually consume rather than on the basis of an average mix of resources.
- Use price differentials to motivate desirable user behavior (such as discounting nighttime prices to shift work to a time when idle capacity is available).
- Employ output-related charges when a reasonably standard measure of output exists (e.g., number of credit accounts maintained or invoices issued) rather than input-related charges (e.g., CPU seconds, input/output volume).
- Charge fixed prices for services that have a relatively small (or predictable) demand for resources, such as a fixed monthly price for dedicated workstation and electronic mail services.
- Use a fixed-price contract for each stage of a software development project, possibly with a fixed upper limit on the project as a whole.

- For large development projects that could have a significant impact on a user's financial performance during the current accounting period, consider amortizing the development cost over multiple periods so as not to discourage desirable long-term investments.
- Develop a reporting system tied to the chargeback system that gives users the information they need to made intelligent decisions regarding their use of information services.
- Adhere to an "open book" policy that permits any user to ascertain the basis of all charges.

Developing a rational chargeback scheme is a large undertaking. Its design raises some difficult economic, technical, and organizational issues. In some important respects, through, the problem is becoming easier.

In a distributed environment, a growing fraction of computing services is provided on dedicated facilities, which eliminates much of the complexity of record keeping and cost allocations. Furthermore, as core services such as personal workstations and electronic mail become universally used throughout the organization, many charges can be based on a simple head count measure. The vast majority of users consume a fairly predictable and low level of services, and so charging on the basis of minor differences among them would have an insignificant effect on total costs; in these cases, then, charges can be based on average per capita usage of the standard set of services. This leaves only the relatively few users with large discretionary demands for shared resources as the prime target of the more complex pricing schemes.

Designers must avoid a fruitless quest for the perfect chargeback system, because it does not exist. Any system will inevitably call for a host of subjective judgments and approximations. The real test is thoroughly pragmatic: does the system increase the effectiveness and efficiency of the organization's information system? By this test, most chargeback systems could stand considerable improvement.

Further Readings

Bernard, D., J. C. Emery, R. L. Nolan, and R. N. Scott, *Charging for Computer Services: Principles and Guidelines,* Petrocelli, 1977. A good conceptual discussion of the issues in setting chargeback policies, but without much coverage of detailed procedures.

Emery, James C., "Cost/Benefit Analysis of Information Systems," in J. Daniel Couger and Robert W. Knapp, *Systems Analysis Techniques,* John Wiley & Sons, 1974, pp. 395–425. Similar orientation to that given in Chapter 8, but with some additional material.

❖ 9 ❖

Systems Concepts

Despite the huge investments they have made in information systems over the past three decades, most organizations have given surprisingly scant thought to developing an underlying rationale for their MIS. Most MIS managers have been so fully occupied with the struggle to keep their heads above water in the face of explosive changes in the technology that they have had little energy left over to consider the conceptual underpinnings of their efforts. General management, for the most part, has viewed the MIS as a back office matter without much relevance to the basic operation of the enterprise—and certainly not worthy of much deep conceptual thought.

Systems concepts provide valuable insights into designing an effective information system, as well as dealing with the closely related issues of organizational structure, centralization versus decentralization, and management incentives. An organization that aspires to gain strategic advantages from its MIS needs a sound conceptual foundation to guide it in making some very difficult choices. The body of knowledge dealing with systems concepts provides such a framework. It is worthwhile to review some of its major ideas, to apply them to the issues of developing an MIS strategy.

What Is a System?

A **system** is an entity composed of related parts directed at a purposeful activity. Here we are interested in man-made systems designed to accomplish a given set of goals (rather than a biological organism, say, whose creation and goals are best left for others to ponder). More specifically, our focus will be on two types of systems: the organization and the MIS.

Hierarchical Structure

Like all systems, the organization and the MIS are composed of parts, called **subsystems**. As we have already discussed, these systems are broken down into parts because they are too complex to deal with in a monolithic way. Each subsystem is established to handle a portion of the system's activities as a way of simplifying the design and management of the entity as a whole. The factoring of tasks into subtasks typically continues for several levels, in hierarchical fashion. This process proceeds until the resulting pieces of the organization are simple enough to manage without further division.

System Boundary and Environment

A system has a **boundary** that defines the activities considered to be integral parts of the system. In the case of an organization, for example, the boundary would typically encompass the set of activities conducted within the legal corporate entity. The boundary of an MIS may be defined to include all the centralized data processing activities within an organization.

Everything not included within the boundary of a system constitutes its **environment**. The environment of an organization includes, for example, customers, suppliers, competitors, and government agencies. The environment of the MIS includes the people, organizational units, and activities not directly involved in data processing.

The boundary of a system (and hence its environment) is defined essentially arbitrarily for the purpose at hand. The system should include those activities over which a decision maker exerts significant control, leaving the remaining activities to form the environment. Degree of control is, of course, a relative matter. A president can legitimately consider the entire organization as a system, whereas a plant manager would likely view those activities of the organization outside of the confines of the plant as part of his environment.

Although the boundary of a system may be arbitrarily defined, the issue is by no means unimportant. Consider the case of an external supplier. If an arms-length relationship through a competitive marketplace provides the only link with the supplier, then the supplier should certainly be considered part of the environment. Suppose, however, that a long-term contractual relationship exists in which scheduling decisions between the two firms are closely coordinated. Under these circumstances it makes sense to include the external supplier as an integral part of the logistics system of the buying firm, despite the fact that the two firms are not part of the same legal entity. The broader boundary in this instance allows scheduling decisions within the buying firm to pay explicit attention to such factors as capacity limitations and delivery times of the supplier.

A system has **inputs** and **outputs**. An input is obtained from the environment, and an output is provided to the environment. The inputs of a manufacturing firm, for example, include human resources, plant and equipment, raw materials, energy, revenue, and information (sales orders, say); its outputs include finished products and payments for the various input resources (e.g., wages, payments to suppliers, dividends).

A system is designed to accomplish certain purposes. The typical private firm, for example, seeks to make a profit, but its objectives also include other matters having a tenuous or even conflicting relationship to profits (such as the well-being of its employees or the firm's social contributions). The behavior of the organization is governed by a **planning system** that makes strategic, tactical, and operational decisions directed at meeting the firm's objectives. A **control system** provides **feedback** information concerning the actual accomplishment of the organization's plans, which serves as a basis for taking corrective actions when significant deviations occur.

Interactions Among Subsystems

Man-made systems can be enormously complex. A large multinational firm, for example, may have many thousands of employees spread around the globe and engaged in a huge variety of related activities. Managing such an organization is a mind-boggling task.

A large system of this sort is manageable only because it is split into separate subsystems, each simpler than the system as a whole. If each of the subsystems were entirely independent, managing the system would merely require dealing separately with all its constituent parts. Clearly, though, life is much more complicated than this. Much of the complexity of a system stems from the fact that the parts are *not* independent; they mutually affect one another through a variety of **interactions** that cross subsystem boundaries.

Sources of Interactions

Interactions stem from two basic sources: **coupling** and **shared resources**. As shown in Figure 9-1A, Subsystem A is coupled with Subsystem B by virtue of the fact that the output of A serves as an input to B. Any change in A's output rate immediately affects B's input rate, and vice versa.

In Figure 9-1B, Subsystems B and C are both coupled to Subsystem A, drawing upon the same resource (an intermediate raw material in a chemical processing plant, say). Subsystems

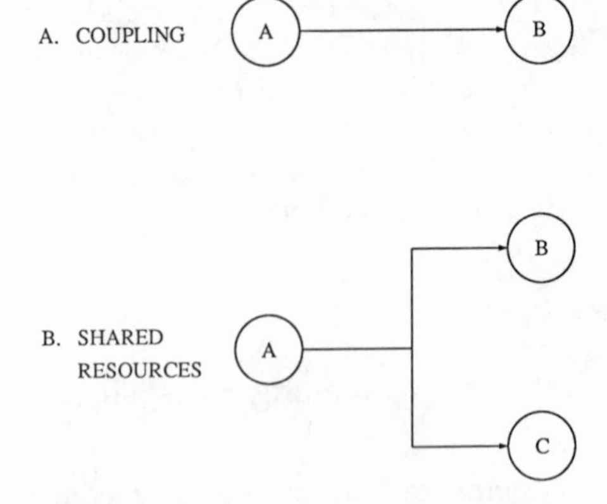

Figure 9-1. Sources of interactions among subsystems. (A) Interaction through coupling. (B) Interaction through shared resources.

B and C thus interact with one another, since any consumption of the common resource by one of them reduces the resources's availability for the other (assuming that A's capacity is not sufficient to meet any combination of demands coming jointly from B and C). Note in particular that B and C interact even though they themselves are not coupled.

The existence of interactions adds greatly to the complexity of managing a system. A change made in one subsystem tends to have ripple effects in other subsystems. As a consequence, decision makers must generally pay attention to factors outside of their immediate view.

Reducing Interactions Through Choice of Structure

A variety of approaches can be used to deal with these interactions. The most basic one is to eliminate or reduce them as much as possible through choice of the system structure.

The way in which an activity is factored into subtasks has a major effect on the degree of interaction among the subtasks. Figure 9-2 shows an abbreviated version of two example structures for a manufacturing firm. Structure A factors the top level of the organization into the functions of marketing, engineering, and manufacturing. Each function, in turn, is further broken down into a consumer products department and an industrial products department. Structure B, in contrast, factors the organization into two product groups, and then breaks down each group by function.

Which structure best reduces interactions? It is impossible to say without looking into the underlying sources of the interactions. Suppose, though, that consumer and industrial products have very little in common—for example, they are based on different technologies, produced with different processes in different manufacturing facilities, and distributed through different marketing channels. Thus, very few interactions exist between the two independent businesses of the firm.

The product-oriented structure, Figure 9-2B, certainly makes more sense under these circumstances, since almost all the interactions can be dealt with entirely within each product group. The few interactions that cross product group boundaries (arising primarily from the allocation of the firm's financial resources) add relatively little extra complexity. The functional structure, in contrast, would require the handling of most interactions at the top level in the organization.

One should not conclude from this simple example, of course, that a product-oriented structure enjoys an intrinsic advantage over a functional structure; its advantage in this particular case comes only from the nature of the interactions involved in the situation. If consumer and industrial products shared (or potentially could share) technology, production facilities, and distribution channels, then the best structure might very well shift to the functional organization. The only conclusion that can be drawn from these examples is that designers of a system should base its structure on the specific nature of the interactions among the tasks performed within the system.

This is easier said than done. One of the chief difficulties is that a structure that reduces problems of coupling may exacer-

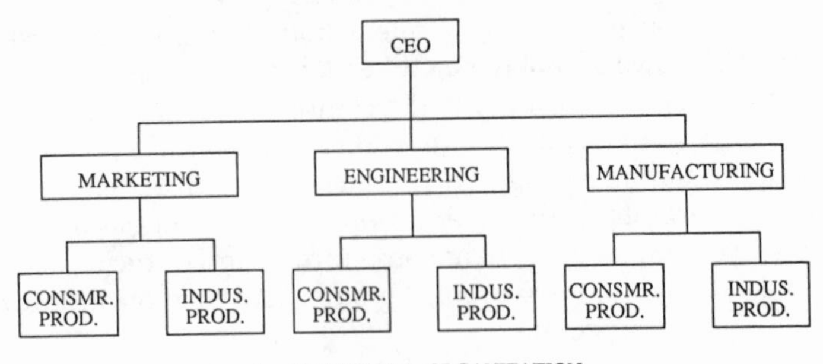

A. FUNCTIONAL ORGANIZATION

VERSUS

CEO

CONSUMER PRODUCTS

INDUSTRIAL PRODUCTS

MARKTNG. ENG. MFG. MARKTNG. ENG. MFG.

B. PRODUCT ORGANIZATION

Figure 9-2. Two alternative structures of a manufacturing firm. (A) Functional organizational structure. (B) Product organizational structure.

bate problems of resource sharing (and vice versa). This is clearly illustrated by the classical issue of functional versus project structure.

A research laboratory, for example, may include a variety of functional skills, such as electrical engineering, mechanical engineering, chemistry, mathematics, economics, and computer science. A laboratory typically conducts multiple simultaneous projects that require a mix of the functional skills. How should the activities be organized?

A **functional structure**, with a manager responsible for each of the separate functions, is one approach. This offers the advantage of dealing well with the acquisition and sharing of the input resources of the laboratory— primarily skilled personnel. On the other hand, it places a severe burden on the managers in dealing with interactions that cross functional boundaries on a multifunctional project.

A **project structure** facilitates interdisciplinary coordination. Each project manager is assigned a staff having the mix of skills appropriate to the work involved. It is then the responsibility of the project manager to make detailed work assignments for this staff. Such a structure deals well with interfunctional tasks, since the coordination does not cross a project boundary. Unfortunately, however, it inhibits efficient sharing of resources across projects, which would involve interdepartmental interactions.

The dual requirement of facilitating efficient resource sharing while managing the coupling within a project is explicitly recognized in the **matrix** form of organization. With this structure, a functional manager typically deals with the hiring, development, and intermediate-term assignment of personnel to individual projects. A project manager must then provide detailed supervision of the staff currently assigned to the project. Each staff member thus looks to two different bosses—one for longer-term development and project assignments, and the other for day-to-day supervision.

Under the proper circumstances, the hybrid matrix structure provides a good compromise between a pure functional or project structure. A matrix structure deals adequately with longer-term resource sharing, while limiting the intensity of day-to-day interactions that cross functional boundaries. Coordination necessary to deal with the shared responsibilities intrinsic to a matrix structure can be managed without great difficulty in an environment of good will, but breaks down quickly in a politicized climate.

Reducing Interactions Through Decoupling

The tradeoffs involved in a functional versus project structure illustrate a general principle: no amount of cleverness on the part of a designer can entirely eliminate interactions that cross subsystem boundaries. After choosing the best structure possible, we must still deal with the remaining interactions.

It should be noted that the importance and complexity of an interaction are issues of degree. An interaction that occurs with great frequency and pervasiveness is obviously of greater concern than one that occurs infrequently or involves relatively minor stakes. Anything that reduces the frequency, detail, or intensity of an interaction makes the management task that much easier.

The effects of coupling between subsystems can be reduced with various forms of **decoupling**. All forms of decoupling aim at increasing the degree of isolation of a subsystem so that it can be managed without having to rely on detailed and frequent knowledge of other parts of the system.

Consider the use of a **buffer**, for example. A buffer is a storage device inserted between a supplying and using subsystem. The supplying subsystem feeds its output to the buffer, rather than directly into the using subsystem; the using subsystem then obtains its input from the buffer. In a chemical plant, for instance, an intermediate raw material is almost always kept temporarily in a storage facility before being consumed in a downstream process. In an on-line information system, messages arriving from remote terminals are moved temporarily into buffer storage before being processed by the computer.

A buffer offers the great advantage of increasing the degree of isolation of the coupled subsystems. In the absence of a buffer, any change in the supplier's output rate or user's input rate immediately impacts the other party. With a buffer, a temporary mismatch between the rates can be absorbed within the limits of the buffer's available capacity. The effects of coupling are felt only when the buffer fills up or runs dry, and hence a tradeoff exists between the size of the buffer stock and the degree of interaction.

The reduction in interactions provided by a buffer depends not only on the size of the buffer, but also on the flexibility of the rate of output. If the supplying subsystem must operate near its maximum output rate to match the consumption rate, any variation in output eventually gets transmitted to the using subsystem. If, on the other hand, **slack capacity** exists, the output rate can be increased to make up for any temporary shortfall.

Standardization provides still another form of decoupling. The supplying subsystem is expected to control the characteristics of its output so that they fall within certain standardized limits. An intermediate raw material, for example, is specified with respect to its chemical composition and the type and amount of allowed impurities. In this way, the using unit can carry on its activities without close coordination with its supplier. Lacking such standardization, the using organization would have to adjust its process in response to variations in the characteristics of incoming material. In the implementation of an information system, standardized interfaces are especially important because they allow developers to work on individual modules without constant and detailed coordination with other development teams.

Managing Interactions Arising from Resource Sharing

Similar approaches apply to dealing with interactions coming from resource sharing. Here, too, slack resources play an important role by mitigating the effect of sharing a scarce resource. There is such a thing as running too tight a ship.

The allocation of a shared resource becomes more difficult as the supply becomes more constrained, increasing the likelihood that each user has to bear the effects of a shortage. Consequently, excess capacity, set at a level that is rarely exceeded by the combined demands for the shared resource, can substantially reduce the degree of interaction (though at the cost of the some idle capacity, to be sure).

Managing Interactions Through Coordination

All the approaches discussed so far for dealing with interactions—structure, decoupling, and slack resources—aim at reducing the volume and frequency of coordination across subsystem boundaries. We must now look at the other side of the coin, and consider the role of coordination in managing a system.

Coordination involves the exchange of information to make decisions that take account of conditions outside the immediate subsystem boundaries. Suppose that A and B are coupled processing units in a plant. The following information is required to coordinate their activities:

- The current status of each unit (e.g., its current inventory level and rate of production).
- A "model" of each unit so that the decision maker can predict the effects of alternative courses of action.
- An "objective function" for each unit to provide a basis for evaluating alternatives.

The model and objective function are not generally formalized in a decision support model; they may merely exist in the mind of human decision maker. Without having some notion of the consequences of alternative actions, however, the decision maker has no rational basis for choosing an action and therefore no basis for coordination.

Coordination is a matter of degree. At one extreme, an automated process control system could maintain extremely detailed models of the processes being coordinated, getting into such matters as chemical reactions, product yields, and the material and energy balance of each process. It could also update status information frequently so that measurements of pressure, temperature, and flow rates are kept current within a matter of a second or so. On the basis of this information, the computer could set the operating variables to optimize the process.

With this degree of coordination, the two units can hardly

be viewed as independent. They are managed as a single entity, with full information about both units, and with a single goal. Thus, we can regard an activity managed as a single integrated entity as an extreme case of close coordination. Such close coordination might be justified in situations involving very strong interactions among the parts of a system.

At the other end of the spectrum, we can consider the case of no coordination. Rather than take action on the basis of combined knowledge of the interacting parts, each unit could manage entirely on its own. If one unit impacts the other, each individual manager is expected to respond in the best way possible based on local information and objectives.

This approach is by no means uncommon. If the degree of interaction between two activities is very weak, it makes sense to ignore the interaction and suffer the (minor) penalty caused by the oversight; in such a case, the cost of coordination could be far more than the penalty avoided. If, for example, the Turbine Department of General Electric buys a light bulb, it need not coordinate with GE's Lamp Division; the Lamp Division can easily respond to the demand through the marketplace, without any direct contact.

Coordination as normally practiced falls between the two extremes. That still leaves a great deal of latitude in the design of the coordinating mechanisms, however. Consider the case of a multiplant firm in which shipments among plants are coordinated at the corporate level. The coordination might take one of several forms:

1. Using aggregate cost and capacity information available at headquarters, monthly interplant shipments are fixed in advance, with no attempt to specify the timing of the shipments (thus requiring the using unit to draw upon buffer stocks during the month).
2. A weekly schedule of shipments is sent to each plant, with the amount shipped on each working day specified in advance; the receiving plants, in turn, are provided a weekly production schedule compatible with the amount of incoming material (and any buffer stocks).
3. Each plant transmits to headquarters a nightly report of the inventory levels and capacities in each of its production facili-

ties; based on a detailed model of each plant's production process and costs, headquarters then returns to each plant a schedule of production and interplant shipments for execution the next day.

The examples given, from no coordination at all to real-time process control, demonstrate the range of coordination that may be called for in different situations. The degree of coordination can be defined in terms of such factors as the following:

- The number of variables used in the coordination (e.g., an aggregate tonnage measure of a plant's capacity versus a detailed capacity measure of its individual work centers).
- The time resolution of the variables (e.g., weekly versus daily output variables).
- The accuracy of the "model" of each unit being coordinated (e.g., a simple-minded cost and capacity model revised each budget cycle or a detailed linear programming model with numerous variables and constraints).
- The frequency with which information is updated and activities are coordinated (e.g., monthly or daily).

Integration vs. Independence

Integration

A fundamental issue in designing a system is the balance between integration and independence. An **integrated** system is one with a high degree of coordination. Inputs and outputs are tightly scheduled, taking into account the effects of one subsystem on the others. Resources are shared widely, with close coordination to allocate capacity among competing demands.

The motivation for such integration is greater overall efficiency and more effective meshing of activities among subsystems. Close coordination permits a reduction in the size of buffers and the amount of slack resources, thus providing greater resource utilization. Extensive resource sharing offers potential benefits from economies of scale and specialization. With coordination, decisions can be made from the perspective of the system as a whole, rather than on a suboptimal basis using only local information and objectives.

Integration comes at the price of added complexity and risk, however. A system is factored into subsystems to avoid complexity, while integration gives up some of this advantage by partially recombining the components. Close integration requires frequent and detailed communication among components of the system. Decision processes that consider the global effects of alternative actions are much more complicated than local suboptimizing processes. An integrated system inherently demands considerable centralization in decision making and control, with attendant bureaucratic penalties of rigidity and delay.

A more subtle, but very real, cost of integration is the risk of doing it badly. In principle integration brings greater efficiency and a global perspective, but in practice the task may be too complex to carry out successfully. Trying to devise and operate a completely integrated and detailed logistics system for a multinational firm, for example, would be a formidable task— one well beyond current practice. An attempt to impose detailed central decisions and constraints on the entire firm, in the hope of gaining the advantages of integration, would almost certainly end up with a bureaucratic, inflexible, and unresponsive system.

Close integration runs other risks as well. A system with tight coupling and extensive resource sharing may be extremely efficient if all goes according to plan, but it is vulnerable to uncertainties. A highly automated, specialized plant may be very efficient in producing a company's total requirements for a critical component (automatic transmissions for General Motors, for example), but it exposes the company to the risk of a serious problem if the single facility fails. Fragmentation has its price, but without some degree of duplication and slack the risk of total failure may be unacceptably high.

Independence

The arguments for and against greater independence mirror those concerned with greater integration. Independence brings simplicity, responsiveness, and robustness in the face of an uncertain world. It does so, however, at the price of forgoing opportunities for greater efficiency. Successful achievement of the suboptimal goals of independent subsystems can only approximate the system's global objectives.

Tradeoffs Between Integration and Independence

Designers of a system face difficult tradeoffs between integration and independence. Both have their pros and cons. The best compromise always involves a blend of the two: almost every system builds in buffers and some slack resources, and almost every system engages in some form of coordination across subsystem boundaries. The only issue is the relative balance between integration and independence. The extent to which a given system should favor one or the other depends on its particular characteristics.

Tight control is called for in the face of strong interactions that have a significant effect on the performance of a system. In the automobile industry, for example, substantial economies of scale can be achieved by sharing production capacity among a variety of products. "Just in time" inventory control, achieved through close coordination with suppliers, can produce major reductions in inventory with concomitant reductions in working capital, floor space, and material handling. With such large potential savings from tight coordination, the tradeoffs favor a relatively high degree of centralization.

Since tight control always exacts its price in complexity and vulnerability to uncertainties, it is warranted only if it provides a means to deal more effectively with significant coupling or resource sharing. In the absence of such benefits, a higher degree of independence is appropriate. Given the natural human tendency of those in charge to overstate the advantages of centralization and understate its difficulties, the burden of proof should probably rest with the centralists.

In a diversified company with weak interactions across its major divisions, one would expect to find at the top level a relatively high degree of decentralization. Independence is especially justified in a situation calling for creative solutions and quick responses in an uncertain environment, such as faced by a high-tech subsidiary that must produce a quick succession of new products in a rapidly changing market.

The issue of integration versus independence recurs at each level in a multilevel system. Decentralization *across* the major groups of the diversified firm may be entirely appropriate in the absence of much coupling or resource sharing at the corporate level. Strong interactions *within* a group, however, could justify a high degree of internal centralization. At every level, a clear rationale should exist for establishing a given combination of coordination and independence.

Effects of Advances in Information Technology

The tradeoff between integration and independence depends on the costs versus the benefits of coordination. Coordination involves communications and information processing, and so advances in technology that lower the cost of information processing often tend to favor a higher degree of integration. In some cases, though, technology may foster greater independence by lowering the benefits of resource sharing. These effects can be seen in the design of both the MIS and the organization's planning and control system.

Design of the Information System

Technological advances can have mixed effects on the issue of integration versus independence of the MIS. Advances in communications and data management technology allow widely dispersed users to access a central data hub tied to a pervasive communications network. Large-scale mainframe processors and powerful development tools make it feasible to deal with the added complexity of applications that benefit from tight coupling and high-volume data sharing—often the heavy-duty mainstream applications that lie at the heart of the enterprise. Little on the horizon suggests that this integrating role of the mainframe will not continue (although some vendors might dispute this assessment).

While these developments favor integration, others favor independence. The powerful, low-cost mini- and microcomputer allow economical distributed computing and the assumption by users of a greater role in meeting their own information needs. Armed with such a computer, along with user-friendly languages, a user can work effectively with relatively little coordination from the central staff. Under these circumstances, sharing the central mainframe and technical staff becomes less attractive. As in the case of other systems, the trick in designing an effective distributed system is to strike the right balance between integration and independence, centralization and decentralization. This is a challenging task.

Design of a Planning and Control System

A relatively primitive information technology makes it difficult to achieve a high degree of integration across the myriad activities of a complex organization. If communications, data collection, data storage, and computation are expensive, slow, and unreliable, then the design of a planning and control system must reflect the organization's limited ability to coordinate the activities of its subunits. The system does this by granting subunits considerable independence so that they can carry on their activities without having to exchange information constantly with a central coordinating authority. Opportunities for the economies that come from resource sharing and close coupling must be forgone because of the difficulties of handling the integration.

Advances in information technology change the economics of integration. Large quantities of coordinating data can be handled rapidly, accurately, and economically. Planning models can be used to aid in the assimilation and use of these large databases. The resulting closer coordination can pay off in lower slack resources, smaller buffer stocks, economies of scale and specialization, and quicker response to events that demand the coordinated action of multiple subunits.

Evidence of these changes appears in a variety of contexts. Military command and control systems, for example, used to grant almost complete autonomy to a theater commander (the British commander, General Howe, during the American Revolutionary War, for example). Now it is not unknown to manage distant squadron movements from the Pentagon.

The logistics operations of multinational firms manifest a similar movement toward closer integration. The growing use of interactive systems creates at each production location a detailed database of operational data dealing with such matters as production output, resource utilization, and raw material costs. These data can then be made available for centralized planning and control purposes, at a relatively low incremental cost. Transmitting daily production figures to a central planning office, for example, would be entirely feasible in most organizations.

As a result of these new capabilities, some companies have found it attractive to achieve a higher degree of specialization in production. Individual components might be produced in different countries, for example, with the assembly and distribution of finished products closely coordinated on a worldwide basis. Scheduling of production can take account of such matters as the varying labor and raw material costs, available capacities, relative efficiencies, and tax effects. In the past, without the support of an advanced coordinating infrastructure, the tradeoffs tended to favor greater self-sufficiency in each national market.

As always, there are tradeoffs involved in a shift to closer integration. Tighter global control has its costs and risks. In assessing the tradeoffs, however, information technology must certainly be taken into account. It clearly provides a means of managing a much higher degree of integration than was technically feasible only a few years ago. It is up to management to apply this technology wisely in seeking the best blend of integration and independence.

Further Readings

Ackoff, Russell L., *Creating the Corporate Future,* John Wiley & Sons, 1981. A provocative and interesting discussion that provides some useful insights about how organizations should design planning and control systems.

Lawler, Edward E. and John Grant Rhode, *Information and Control in Organizations,* Goodyear Publishing Company, 1976. A conceptual discussion of the planning and control process in organizations.

Miller, James G., *Living Systems,* John Wiley & Sons, 1978. A monumental scholarly work dealing with the application of system concepts to a wide variety of systems (organizations, biological, etc.).

Scheer, A.-W., Computer: *A Challenge for Business Administration,* Springer Verlag, 1985. An excellent coverage of systems concepts applied to the design of integrated information systems.

Weinberg, Gerald M., *An Introduction to General Systems Thinking,* John Wiley & Sons, 1975. Systems concepts from the viewpoint of a behavioral scientist, with special relevance for dealing with the design of human interfaces.

❖ 10 ❖

Foundations of a Successful MIS Strategy

After years of considerable indifference, management increasingly regards the organization's information system as a source of competitive advantage, or possibly as a potential threat in the hands of an aggressive competitor. The stakes grow each year, not only due to the swelling investment in information systems, but also because of the high opportunity costs of failing to exploit the technology to gain organizational effectiveness. Under these circumstances, it behooves managers—those with general interests as well as those with specialized MIS responsibilities—to pay careful attention to how the information system can better contribute to achieving the organization's strategic objectives.

MIS Planning

The Need for a Plan

Development of a strategic **MIS plan** provides an opportunity for the organization to consider explicitly how it should exploit the growing capabilities of information technology. If done right, such a plan should provide the following:

- A description of a desired future MIS capability required to support the strategic needs of the organization.
- Guidance for *current* actions aimed at achieving the plan.
- A focus for organized problem solving that draws on a broad set of the talents within the enterprise.
- A vehicle for communication and coordination among involved parties.

Planning is one of those activities that almost everyone supports in principle but frequently shuns in practice. Planning in too many organizations is little more than an autumnal ritual, the results of which get promptly relegated to a high shelf until the next year. A useful plan must provide guidance for making relatively short-term decisions on such matters as the allocation of resources and the selection among technical alternatives. A plan that has no bearing on these decisions is too future-oriented or too irrelevant to contribute much value.

The MIS plan must also take a longer-term view, because it takes a long series of purposeful actions to put into place the kind of system that can support the firm's strategic objectives. A technical staff must be hired and trained to provide the necessary mix of skills compatible with the plan. An enabling technical **infrastructure**—dealing principally with communications and the management of data resources—must be established as a base for the development of future applications. Priorities must be set on application development projects. Without a broad picture of the firm's long-term business objectives and its supporting information needs, it is very difficult to make intelligent choices along the way.

The Context for MIS Planning

As the MIS becomes more and more embedded in organizational activities, almost everything has relevance for the design of the information system. The organizational structure, for example, can have a significant effect on the structure of the MIS (and vice versa). Many applications are implemented within existing organizational boundaries. This approach can simplify client relationships, but possibly at a high cost in forgone opportunities for improvements. With the advent of a technology that facilitates communications and data sharing, it becomes increasingly attractive to integrate the firm's mainline applications—that is, those that lie at the heart of its business—and these often cross existing organizational boundaries.

In a firm that manufactures high-tech industrial products, for example, the mainline activities would typically include order entry, engineering design, production scheduling, inventory control, and distribution. These functions would almost always overlap existing organizational boundaries (e.g., Marketing, Manufacturing, and Engineering). A similar overlap occurs in a college that sets out to develop such "student lifeline" applications as admissions, registration, housing, student billing, and alumni relations. In both these examples, existing technology is likely to favor a much higher degree of integration than previously obtained; and in both cases, rigid adherence to existing organizational boundaries would surely result in a system that falls short of meeting the long-term needs of the organization.

As we saw in the previous chapter, information technology is likely to have apparently conflicting effects on the issue of centralization versus decentralization. On the one hand, sophisticated communications and data sharing facilitate centralization in cases where closer coordination brings substantial benefits. On the other hand, low-cost hardware and user-friendly software tend to push greater responsibilities down toward the direct user. A distributed system should aim at a hybrid approach that combines the best features of centralization and decentralization.

An organization's **incentive** and reward system impacts the design of the MIS. The MIS provides the primary means for measuring and reporting performance. What is measured, and what is reported, has (by intent) behavioral effects on managers throughout the organization. It is critical, therefore, that the measurement and reporting system recognize these effects in order to steer behavior toward the interests of the organization as a whole.

At a somewhat more detailed level, the MIS must be designed to support the specific business plans of the organization. In a bank or insurance company, for example, almost any product innovation has important implications in terms of the supporting information technology. Organizations in other sectors also find that their MIS increasingly plays a critical a role in implementing changes in products or services.

In bygone days when information technology stood in a less pivotal position, the MIS was expected to adapt to changes in organizational structure, task assignments, incentives, measurement and reporting requirements, and detailed business plans. Now the direction of adaptation may sometimes work in the opposite direction, with changes in information technology motivating and enabling—or even forcing—changes in structure, responsibilities, and business planning. As an obvious example, an organization that does not consider the profound effects of contemporary communications technology on the way it runs its business is at best missing some useful opportunities and at worst jeopardizing its very existence.

That the information system must not be designed in a vacuum should come as no surprise to anyone who understands the fundamental role of information in running an organization. What is surprising, though, is how poorly many organizations manage the links between general strategic planning and their information system. In relatively few organizations, for example, does the senior MIS executive participate in top-level planning councils—due, unfortunately, to a common perception that the MIS official cannot rise above a narrow technical perspective. Yet, without an effective bridge between business and MIS planning, it is extremely difficult for an organization to fully exploit information.

Creating a Shared MIS Vision

Lack of a shared vision inhibits the strategic use of information technology. The technology has the potential to bring about profound changes, but success will be elusive without some

sion. For those that have not, the fault lies not in their stars but in themselves.

Barriers to a Shared Vision

An inhibitor to a wider acceptance of an ambitious vision is management's reluctance to embark on a lengthy and uncertain path. Managers quite realistically see no easy solutions, and may harbor serious reservations about their organization's commitment and ability to pursue continual improvements in the MIS over an extended period of time. Under these circumstances, management can easily give in to the temptation to put off the quest.

The remedy is to focus on the next steps rather than on the end result. A long-term goal is necessary to point the direction and establish a charter, but the payoff comes from a succession of cost-effective applications along the way. Short-term benefits make it much easier to justify the long-term effort. In the process, organizational learning takes place and the vision evolves.

Interrelations with the MIS

As we have seen, the design of an information system is related to a number of broad issues not ordinarily viewed as the province of the MIS staff. Behavior of the organization depends on the mutual relationships among its structure, planning and control processes, incentives, and measurement and reporting systems. As the "central nervous system" of the organization, the information system ties together these governance mechanisms. No fundamental improvements can come through information technology without considering the broader aspects of organizational behavior.

Example: A Retail Bank

Let us look at the example of a retail bank to examine the relationship of the information system to the broader aspects of its business strategy. The bank's major lines of business—checking accounts, savings accounts, trusts, mortgages, consumer loans, and credit cards—historically were managed by separate departments. Each line had its own customer account numbering system with no cross references among them, which made it very difficult to exchange data among the accounts of a given customer. As a result, the customer saw each line as a separate, uncoordinated service. Tasks that involved more than one account, such as a loan repayment from the checking account, required multiple unrelated transactions.

The obvious disadvantages of such fragmentation could only be remedied with advances in information technology permitting the bank to establish links across lines of business (using a common account number when feasible). This allowed the bank to reorganize its contact with customers so that a single representative could handle all aspects of a customer's business with the bank. The changes in technology and organizational structure led to better customer service, a more attractive job for the bank representative, and, in the longer run, the opportunity to develop a variety of complementary financial services that build on a multifaceted relationship with a customer.

Example: A Manufacturing Firm

Now take the example of an international firm producing a variety of industrial products. The development of an integrated logistics system has had some major effects on the way the company does business. Its tightly coupled mainline applications—order entry, inventory control, production control, engineering, and distribution—integrate functions that were previously managed in a fairly fragmented fashion by Sales, Manufacturing, and Engineering. Rapid communication with its widely scattered plants allows a centralized scheduling group to shift production to the currently most economical location and to take advantage of idle capacity and the economies offered by a plant's specialized facilities.

Planning and control within the firm rely heavily on the information system. The detailed scheduling system at the plant level implements a centrally determined aggregate schedule. The system allows operational management to test alternatives and choose the one that best meets the plant's objectives. Feedback data from the operational system provide control to measure deviations from the current schedule; significant deviations are then reported to plant management and, if serious enough to affect the corporate-wide schedule, to the central scheduling group.

The operational planning and control system has a number of integrated links with higher-level forms of planning. For example, the models used for corporate and plant scheduling can also be used to test alternative plans dealing with such matters as plant capacity, inventory stocking policies, distribution patterns, make-buy decisions, annual budgeting, and corporate financing.

A backbone planning and control system such as this provides the basis for a more rational incentive system. It allows performance of the sales department, for example, to be measured by its incremental profit contribution as determined by the scheduling system. The alternatives of using, say, the average contribution over a broad product group, or—even worse—sales revenue, fails to recognize such things as current costs and capacity constraints; as a result, these measurements might motivate dysfunctional behavior on the part of the sales force by encouraging it to emphasize low-profit (but high-commission) products.

The information system's detailed involvement in activities at the operational and tactical level—and, to a lesser extent, at the strategic level—gives the organization an unprecedented degree of control. The planning system maintains information about plans, and the transaction processing system collects data that show the actual accomplishments against those plans. The information system therefore has the ability to measure and report on a wide variety of significant matters involved in the internal operation of the firm.

Measurement and Reporting

A sophisticated MIS gives management a very powerful tool, but one that must be wielded with great caution and restraint. Behavior is influenced by the mere act of measuring and reporting it—and not necessarily in the intended way. In choosing variables to measure, one must assume that those whose performance is being assessed will act in a way that makes the measures look favorable. Ideally, of course, one would want a favorable measure to correlate closely with a favorable achievement of the organization's objectives, but this is easier said than done. In almost any reporting system one can find examples of measures that motivate harmful behavior.

Suppose, for example, that a plant's performance in meeting its schedule commitments is measured by the percentage of the jobs delivered on time. If an uncompleted job is already late, this measure of performance gives no encouragement at all for plant management to complete the job as soon as possible: the damage—in terms of the measure—has already been done and it will not be improved no matter when the job is delivered. To improve apparent performance, management is motivated to divert effort from the late job to one that can be delivered on time, even though the real penalty of additional lateness may be much higher on the former.

What is needed is a more intelligent measure that incorporates judgments about how management really wants the plant to allocate its resources. Such a measure would undoubtedly give some weight to the degree of a job's lateness, rather than penalizing the plant in an abrupt single step. Figure 10-1 shows three possible weighting functions.

The first considers only the percentage of late jobs; the second assesses a constant penalty for each day a job is late, and the third uses a penalty that grows steeper for each day late. A more complex function might take account of factors other

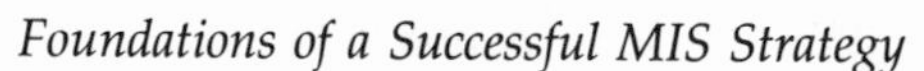

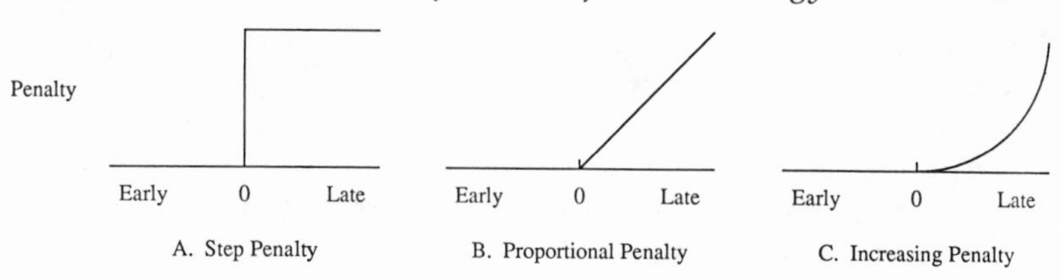

Figure 10-1. Weighting functions for job lateness. (A) Percentage late. (B) Constant daily penalty. (C) Increasing daily penalty.

than lateness, such as the value of a job or the importance of the customer. Any serious attempt to define such a function would certainly have to involve appropriate managers from such groups as manufacturing, sales, and purchasing.

The approach behind most existing reporting systems grew out of an era when information processing was expensive; consequently, these systems tend to minimize the amount of computing involved in the reporting process. With current technology, however, the extra complexity of a sophisticated measurement system can easily be accommodated. The only issue becomes whether it makes sense to try to capture the goals of the organization formally within the reporting system.

In the example of the penalty function for job lateness, one cannot argue that a formal mathematical relationship can capture all the factors an experienced manager would use in assessing the relative importance of job lateness. With hundreds or even thousands of jobs for each reporting period, however, it is impossible to exercise experienced human judgment on each one. To convey a concise picture of an important aspect of the plant's performance, system designers have no choice but to boil down vast quantities of data into a condensed measure. This is particularly necessary in reporting to a senior manger—the corporate VP of Manufacturing, say—who does not have time to personally observe closely all activities for which he or she is responsible. As imperfect as it may be, a weighting function that tries to reflect management's best judgment is likely to be a great deal better than one that is clearly simplistic and potentially mischievous.

Naturally, not all aspects of a plant's performance can be boiled down into a single variable. It is reasonable to seek a composite measure for schedule performance; it would not be reasonable to ask management for a single measure that combines schedule performance with, say, product quality and plant safety. Any such measure would obscure far more than it reveals.

To provide a comprehensive view, it is thus necessary to report on multiple performance variables, each of which is a weighted composite of related factors (such as the lateness of the individual jobs in the shop). The number of such variables should be kept to a practical minimum so as not to swamp the cognitive capacity of the recipient—about half a dozen variables is generally recommended by psychologists and management theorists. For a plant as a whole, separate variables to represent performance with respect to cost, quality, delivery schedules, safety, human resource development, and long-term improvements might suffice to capture the essence of its performance. For a dean at a large university, the system to report departmental performance might use separate measures for enrollments, teaching effectiveness, student quality, research quality, research funding, and cost control.

Any management information system is replete with instances where more sophisticated performance measures would add greatly to the information conveyed to management. These matters receive far too little attention in the design of an MIS. For the most part, simple measures of performance are chosen by default or tradition, often at quite low levels in the organization, without much concern about their implicit policy declarations or the behavior they are likely to induce.

Taking a Comprehensive Approach to MIS Implementation

In principle it is clearly desirable for MIS implementers to take into account the interrelationships among the information system, organizational structure, planning and control, and performance measurements. The difficulty, of course, is that not every application programmer can be concerned with the grand design of the organization before writing a line of code.

An application development team often lacks both the charter and the skills to examine broader issues.

It is nevertheless the case that the implementation of a new information system sometimes triggers a comprehensive review of the organization's basic way of doing business. Those relatively rare situations in which a brand new system is designed from scratch present a particularly good opportunity. In such an event, it would be a gross mistake not to use the occasion to examine and question carefully the basic conduct of the business.

A good system is a simple system. Paradoxically, advanced technology often provides a means to simplify—to make the system "so advanced, it's simple," as the advertising slogan puts it. Many existing systems, designed years ago on the basis of a much less powerful technology, impose burdens on users because programmers could not efficiently handle the functions internally. Lack of integration across files, for example, forces users to enter redundant data and extract related information from multiple reports.

Over the course of many years, a system acquires all sorts of convoluted modifications and extensions to handle special cases or to deal with unanticipated needs. The system eventually loses whatever consistency it might have had. A new design affords the opportunity to clean up the mess.

The redesign should go further than mere "paperwork" simplification; it should consider fundamental practices as well. In the case of one organization, for example, which manages pensions and benefits for a large nonprofit organization, the implementation of a new system brought startling changes in many of its activities. As a result of the study that accompanied the implementation, the following changes were introduced:

- Billing was changed from quarterly payments in arrears to monthly payments in advance.
- A cash management system was instituted that drastically speeded up the deposit of incoming funds.
- Interactive terminals scattered throughout the organization support virtually all its routine work.
- Several new services were made available to participants in the various benefit programs.

- Substantial changes were made in work assignments and management responsibilities.

Some of these changes were anticipated in the initial design, but many evolved out of experience with the system. No amount of up-front needs analysis could have predicted the resulting stream of changes, and only a very flexible system could have accommodated them. The initial system was justified on the basis of expected tangible savings (which were more than realized), but many of the longer-term and fundamental benefits either were not foreseen or were incapable of being expressed in monetary terms (such as major improvements in the quality of service provided to participants).

Most changes would simply not have been feasible with the previous inflexible and fragile system. Other changes *could* have been made without drastic modifications of the old system (such as the changes in the billing cycle), but the fact is that without the impetus of a new system it is very unlikely that management and its board of trustees would have been willing to incur the disruption and risk of piecemeal changes.

Organizing the Central MIS Staff

The Facilitating Role

Putting together a successful information system calls for a high order of leadership on the part of the executive in charge of MIS activities. Rather than playing the more traditional role of computing czar, the MIS executive must be able to thrive in a distributed environment in which everyone gets into the act.

Succinctly put, the dominant emerging role of the central MIS staff is to provide users with a productive environment within which they can meet their information needs. This differs from the conventional function of the central organization as sole supplier of a full range of development and operational services. The new emphasis on facilitation rather than operations makes the job of the MIS staff more challenging and strategic, and generally calls for a new set of skills.

In looking at this new role, one should determine the kind of environment that users would find most productive in meeting their information needs. The following characteristics serve as goals toward which the central staff should strive:

- The ability to connect, easily and efficiently, with any resource on the distributed network.
- The ability to retrieve and update data in a shared database.
- A wide range of supported services, from those associated with the traditional mainframe to those provided through a personal workstation.
- A set of languages and development tools that make it relatively easy for users to satisfy many of their own needs without the direct aid or intervention of the central staff.
- A well-defined standard environment aimed at simplifying interconnection, providing high-quality user support, and shielding the organization from the risks of computerized anarchy.
- A secure environment that offers the means to protect data and other resources from unauthorized access.

Central Services

In the new environment, the core responsibility of the MIS staff is to manage the central infrastructure of the distributed system—the communications network and the shared data resources. Because these functions are fundamental to dealing with coupling across applications and sharing of common resources, they are intrinsically centralized in nature. It is difficult to conceive of a successful distributed system without tightly coordinated communications and data sharing.

Operation of a central mainframe facility is closely tied to the core infrastructure. In serving as a central data hub, the mainframe must provide computational capacity for managing the database. Since it is difficult and expensive to provide tight coupling among applications run on geographically dispersed machines, many large-scale integrated applications can be executed most effectively on a central machine where the shared database resides.

User Support Services

Success in moving responsibility out toward the user community calls for the provision of numerous user support services. Competent support offers a tremendously valuable aid, because a technical task that a trained specialist views as trivial could absorb hours of frustrating and unproductive effort on the part of an inexperienced user. Many of these services can be provided best through the central organization to gain the economies of resource sharing, such as greater technical specialization and the avoidance of undesirable duplication. The following three services are critical to the user-support function:

1. User education.
2. Implementation assistance (feasibility assessment, design, programming, integration with other parts of the system).
3. Selection, acquisition, and maintenance of hardware and system software.

Not all support services should be provided centrally. Services with a relatively low technical content combined with a strong functional orientation, such as financial analysis, should have close spatial and organizational proximity to the user. Specialized application skills do not lend themselves very well to corporate-wide sharing, and therefore little is gained and much is lost by centralizing them. It makes much more sense to manage and budget specialized support personnel at the user level, where it is much easier to build fruitful long-term relationships with users and create strong incentives for providing responsive, cost-effective services.

Standards

The central staff must have responsibility for setting the **standards** defining the environment within which applications can be implemented. Examples of such standards are the following:

Hardware and system software
Programming languages and development tools
Application packages
Data definitions
Communications protocols
Security
Documentation
Cost-benefit justification procedures
Installation procedures

Standards play a vital role in creating a cohesive information system. Without standards it is very difficult to couple one component of the system to any other component, an essential requirement for integration. Sharing a resource implies that it can be allocated among various competing uses; this can only be achieved through standardization. Standardized hardware and software, data definitions, and communication services are key ingredients in providing users with a supportable computing environment. Standards, by definition, apply across the organization as a whole (or at least a major part of it), and therefore the central staff must take the primary responsibility for setting and administering them.

Setting a standard entails some risk. Premature adoption of a standard may lock the organization into an obsolete or dead-end technology. Furthermore, a narrow set of rigidly enforced standards runs a serious danger of imposing an inappropriate tool on users who have unusual needs. Some central staffs tend to give adherence to a standard a purpose of its own, forgetting that the intent of the standard is to facilitate getting useful work done.

A rational approach to managing standards can greatly moderate these difficulties. The following policies serve this end:

- Reduce the risk of obsolescence by delaying the choice of a standard until the long-term winning technology becomes reasonably evident (balanced against the risk of ensuing anarchy until the standard is set).
- Devote the effort necessary to choose a high-quality standard (without engaging in an endless search for the ideal one), so that users can have confidence that a standard will adequately meet their needs.
- Keep the number of standard products to a practical minimum to be able to maintain high-quality support for the selected few products.
- Provide a sufficient range of standard products to meet a variety of needs (such as one word-processing product for the heavy-duty professional user and another for the casual user, since the two needs are quite different and not likely to be met satisfactorily with a single product).
- Permit variance from a standard if a nonstandard product is clearly superior in meeting a given user's special needs (as long as the user is willing to assume the burden of going on his own).
- Be willing to move to a new standard if it offers sufficient advantages to justify the resulting disruptions of conversion.
- Supply bridges when possible between a new standard and the one it replaces to ease the problems of transition (such as providing automatic transfer of documents from the old word processing standard to the new one).

Concept of the Chief Information Officer

As the pervasive and fundamental role of information processing becomes increasingly appreciated, arguments for greater integration gain greater weight. It is difficult to see how any organization can fully exploit the strategic advantages of information technology without creating mechanisms for achieving close coordination among its information-intensive activities. An important aspect of this is the appointment of a high-level executive to run the central MIS function and take the primary staff responsibilities for information-intensive activities throughout the organization.

Examples abound in which the information revolution is beginning to be felt in areas that previously had been quite insulated from the computer:

Office information systems

Communications (data, voice, **images, facsimile,** electronic mail, **voice mail**).

Information services (e.g., the library, public databases, external services accessed through a network).

Reprographics, document processing, and **electronic publishing.**

Process control in manufacturing operations.

CAD/CAM (Computer-Assisted Design/Computer-Assisted Manufacturing).

Organizations are increasingly recognizing the close affinity that these information-intensive activities have for one another. Consider these examples:

- Data stored in a mortgage account (the current balance of a loan, say) are extracted and inserted in a letter to the customer.
- An executive wishing to communicate with a colleague in a distant part of the world can do so through a personal workstation linked to a system that maintains a telephone list of corporate personnel; with the aid of the workstation, the executive can then automatically dial a selected telephone number or can store a voice or typed message in a form that can easily be accessed at the convenience of the recipient.
- Documents produced by the word processing software are distributed through the data communications network.
- Voice and data communications are transmitted in digital form over the same internal network.
- An external bibliographic database is accessed through a writer's personal workstation and incorporated into a research document.
- Data extracted from an external econometric database are incorporated into an analyst's DSS model.
- The company newsletter is composed and laid out on a personal workstation and then printed on an inexpensive laser

printer giving quality approaching that of a specialized (and quite expensive) photocomposition device.
- The production schedule generated by the logistics system is linked automatically to the process control system in a petroleum refinery.
- An order for a custom-designed product is fed from the order entry system to an engineer's CAD system.

Each example illustrates an important potential opportunity for more effective resource sharing or closer coupling of related functions. They therefore make attractive candidates for consolidation under one organizational umbrella. They all exhibit some common characteristics that lend themselves to such integration:

1. Digital representation of information, providing a common means of exchanging data among application areas.
2. Functionally related tasks that can benefit from closer integration.
3. Opportunities to share common resources (communication lines, personal workstations, technical expertise).
4. Similar design and implementation methodologies.

Most organizations have already taken steps to integrate at least some of these activities. The majority of Fortune 500 firms, for example, have placed responsibility for office automation under the head of information systems. This constitutes a considerable break with the traditional separation—even hostility—between office management and data processing. Many organizations are similarly combining voice and data communications, which historically have been quite distinct.

The integration of reprographics, printing, and library services appears to be taking place less rapidly, but is probably inevitable in most organizations with significant activities in these areas. The affinity between business information processing and engineering computation is not widely reflected in the organization structure, but the long-term prospects for closer integration seem quite good. An organization seriously setting out to develop a "factory of the future," for example, would surely have to achieve a high degree of integration among order entry, engineering design, and the robotic devices in the factory.

The generic title for the executive responsible for information systems is **chief information officer**, or **CIO**. Few organizations actually use this title, sticking instead with more traditional titles such as Director of Data Processing or Vice President of Information Services. The job's title, of course, is much less important than its responsibilities. In most organizations the job includes operating responsibility for the centrally managed services (e.g., the network, the central computer center and data hub, and the central technical staff of analysts and programmers). In other matters, the CIO has staff responsibility for overseeing and facilitating the productive use of information technology throughout the enterprise.

In the early days of data processing, the focus in most organizations was heavily on accounting functions. As a consequence, the head of data processing typically reported through the chief financial officer. This is still the most common reporting relationship.

A strategic information system must clearly serve more than the financial functions of the organization. In fact, most strategic opportunities exist in the firm's mainline activities, such as marketing, manufacturing, distribution, and engineering. An undue bias in favor of financial applications can be thoroughly inimical to the broader needs of the enterprise.

Regardless of the executive to whom the corporate MIS group reports, it is critical that the MIS function have a broad charter encouraging a focus on strategic applications (while also getting out the payroll, to be sure). Furthermore, the chief information officer should report at a level consistent with the role's importance and broad responsibilities. He or she usually does not report to the organization's chief executive officer, but reporting at least to the top staff executive (with a title such as Executive Vice President of Administration, for example) makes a good deal of sense in most organizations.

Enabling Technological Capability

In discussing fundamental issues of MIS strategy, one hesitates to dwell too much on the technology. The really difficult parts of achieving success have much more to do with the quality of management than the state of technology. Nevertheless, without an adequate **enabling technology** management would find it impossible to gain much strategic advantage through its information system. Good technology is a necessary if not sufficient requirement for success, and good management must see that it is put into place.

The Problem of Assessing Technology

How can an executive without a strong technical background tell whether the organization uses good technology? A competent systems person can generally make such an assessment, but to whom does the executive look for such counsel? People with apparently equally creditable credentials can differ in fundamental ways. If an experienced data processing professional argues that fourth generation languages are impractical because of their machine inefficiency, how confidently can an executive dispute this assertion? Time and pain may eventually reveal bad past practices, but they do not necessarily point the way to the right current practice.

The internal technical staff may be fully capable of rendering good advice, but it is hardly a disinterested party. Besides, if the staff is not competent and asserts that it is, how is the executive to tell? Use of outside consultants scarcely eliminates the exposure to incompetence and self-interest. The situation is very much like the puzzle of finding the truth when confronting habitual liars and truth tellers (with neither having any obvious distinguishing characteristics): one can ask questions and get answers, but how do you know if you are talking to a liar or truth teller?

The ideal solution is for the executive to acquire sufficient insights about the technology to distinguish good advice from bad. This is probably feasible—I could hardly think otherwise, for that is the purpose of this book—but finding credible sources of help still remains an issue. It is quite likely, for example, that this book contains a number of ideas and assertions that some competent and experienced hands might well regard as unproven, muddleheaded, or flat wrong. Sorting out things in the din of conflicting views is not at all easy. Perhaps the best advice is to remain skeptical of any advice (excluding this), and subject it all to the test of plausibility and empirical results.

Capabilities of an Enabling Technology

The ideal enabling technology sets up a minimum of barriers to accomplishing what the business needs of the organization dictate. In practice, of course, no technology offers instant satisfaction for every need, but some are decidedly better than others in approximating the ideal.

It is reasonable to expect an effective enabling technology to provide the following capabilities:

- Effective **connectivity** through a network with a wide range of devices—computers, workstations, printers, disk drives, and the like (including those outside of the corporate boundary).
- Easy connection with other people, with choices of form (e.g., voice or mail, real-time connection or store-and-retrieve access, text or graphics).
- Easy access to (authorized) data existing in an internal or external database, for retrieval or updating.
- A comprehensive set of tools for implementing applications.
- User support services to facilitate end-user development of applications.

The quality of these capabilities, and the ease with which they are used, can vary greatly. Even in the best of cases, nothing is entirely easy. The first steps are always the hardest: building the base infrastructure of a network, a mechanism for data sharing, and the initial core applications. After that, enhancements and extensions come much more easily.

Building an Enabling Technological Base

Putting in place an appropriate technological base is one of the most demanding parts of building a successful MIS strategy. The task can be challenging and interesting enough to technologists, but most general managers probably view it with a mixture of awe and distaste. It requires hard work that never ends. Nevertheless, it can spell the difference between the organization's success or failure to realize its strategic MIS goals.

It is impossible to put in capsule form all the myriad tasks necessary to build such a base, but certainly the following actions must be taken:

- Develop long-term aggregate estimates (e.g., over three to five years) of capacity requirements, along with the necessary staffing and funding.
- Develop detailed short-term budgets (e.g., over a two-year horizon) for MIS activities within the general guidelines of the longer-term plan.
- Develop means for educating users in MIS concepts.
- Manage procedures for attracting, training, and providing attractive career paths for the technical staff.
- Plan and manage the implementation and operation of a communications network.
- Plan and manage the implementation and operation of data sharing mechanisms (e.g., data dictionary, database management system, data administration).
- Plan and manage the installation of office information systems, with special emphasis on text processing, interpersonal communications, and document transmission.
- Plan and manage the implementation of centrally developed application programs.
- Plan and manage user support services.
- Provide a high-productivity software development environment (e.g., 4GLs, extensive program library, interactive debugging facilities with subsecond response).

- Develop procedures for project planning and control.
- Develop procedures for budgeting and charging for information services, including both user- and centrally funded activities.
- Develop vendor selection policies (e.g., a single- versus multiple-vendor environment).
- Develop software acquisition policies (e.g., make versus buy).
- Develop mechanisms for identifying and introducing promising new technology.

The list could go on—the one given is certainly not exhaustive. Each one of these tasks contributes an important piece to an effective MIS strategy. And each is a formidable undertaking. Putting together a communications network for a large organization, for example, is by itself a mind-boggling chore. It is no wonder that the development of a sound technological base is so difficult and demanding. Despite the effort and resources devoted to this goal, it is far from realized in most organizations.

Setting a Policy for Adopting New Technology

With a technology that changes so rapidly, MIS management faces the never-ending task of constantly renewing its technological base. Many technologies that show early promise lead up blind alleys, or take much longer than predicted to reach a state of practical application. Tracking a technology and developing internal skills in its application can be an expensive business, and therefore backing a dead-end technology carries a real penalty. Not the least of the penalties is the opportunity cost of failing to bet on a winning technology. With these uncertainties and high stakes, the organization needs to develop effective mechanisms for technology assessment.

The technical risks associated with the introduction of new technology can be significantly reduced by waiting until others have gained experience with it. Countering a strategy of delay, however, is the risk that a new and significant technology in the hands of a competitor may jeopardize the laggards. Management should thus give serious thought to setting an appropriate aspiration level for applying new information technology. Any of the following strategies could plausibly be considered:

1. Maintain technological leadership, aspiring to gain a competitive advantage through new technologies.
2. Be an early follower, expecting to catch up relatively easily with any competitor that gains a temporary lead in information technology.
3. Be a casual follower, assuming that the organization can not or need not keep up with information technology (and hence looks to other sources of competitive advantage).

As in most matters, aspirations are a matter of degree. Even an organization that aspires to leadership must select carefully the areas in which it believes such leadership will pay off; even an organization that sets a low aspiration level should not be indifferent to technologies that affect the heart of its business. Modest-sized organizations need not despair of keeping fairly close to the leaders in a few selected areas, because the highly competitive hardware and software market provides an extremely efficient mechanism for rapid and low-cost technological transfer.

It is not easy to make judgments about the technological positioning of the firm. The history of information systems is laden with examples of companies that gained an early technological lead, only to find competitors replicating and often improving upon the technology at a much lower cost and with little damage to their competitive position. The often-cited warning "you can tell the pioneers by the arrows in their backs" attests to the pain that often accompanies technological leadership. The banks that installed early automatic teller machines, for example, paid a high price and reaped few benefits for their pioneering efforts because competitors could react well within the slow adoption time of most consumers.

In other cases, though, competitors have been pushed into a disastrous corner by their failure to keep up with a technological innovator. American Airlines with its Sabre reservation system presents a classic example. The company developed the first large-scale commercial application of an on-line system. It paid an enormous learning cost for its leadership, in the face of a great deal of skepticism about the payoff and an industry-wide refusal to join the early effort. Eventually, after many years of continual improvements and huge investments, Sabre became the heart of the company's business and a major factor in its establishing a strong competitive position. The system reaches out through a communications network to most travel agents, who sell about three-quarters of all airline tickets.

Some competing airlines have been merged or driven into bankruptcy, at least in part because of their lack of a comparable system. They suffered from the barriers imposed by the time and cost of developing their own system. Even when armed with similar capabilities, competitors found it difficult to convince travel agents to install rival terminals after the agents had become closely wedded to the Sabre system.

The contrasting examples of automated teller machines and airline reservation systems demonstrate that there are no easy answers to setting policies on introducing technology. Indifference is surely not justified, but neither is a uniform attempt to keep ahead across a broad front. Effort should be concentrated on technology that stands a decent chance of contributing to the mainline activities of the firm, leaving for others the task of pioneering in the more peripheral areas. A policy declaration to this effect certainly does not resolve the issue of which technologies deserve the firm's attention, but it does place some discipline on the selection.

Mechanisms for Technological Assessment and Deployment

Important technological developments almost always send early signals of their impending arrival. Database management and artificial intelligence, for example, were heralded long before they became fashionable in practice. Unfortunately, though, technologies that eventually prove unattractive often send similarly favorable signs. Separating one from the other is the most difficult part of technology assessment.

Given these difficulties, the assessment and deployment of technology should be managed through a staged process. The process begins with a relatively superficial scanning across a broad range of technologies, and eventually ends by winnowing them down to a few selected for deployment. The following stages and mechanisms seem to work:

- Broad scanning, using such sources as professional journals, trade publications, professional associations, peer groups, public conferences and seminars, supported university research, vendor representatives, and technology-tracking consultants.
- For those technologies that seem most promising, assignment of internal responsibility for keeping on top of them.
- Pilot application of a selected technology.
- Production deployment of applications that successfully survive the assessment of their pilot installations.

A few important points must be made about this staged process. First, we should note that an organization's attitude about technical complexity depends a great deal on its existing technological base. For some organizations the installation of a local area network might stretch their technical skills, while for others it could be entirely routine. Technology assessment aims at infusing new skills relative to the organization's current capabilities.

It should also be noted that the difficulties of assimilating new applications have more than just a technological dimension. The installation of office automation, for example, may introduce little new technology per se, but it may require considerable organizational learning about such matters as the sociology of an office, allocation of office tasks, and proper lighting conditions for efficient word processing. Managing organizational learning of this sort requires much the same process as the staged infusion of new technology.

Faced with so many potentially attractive applications, an organization must employ an efficient screening mechanism capable of identifying significant technologies. The trick is to keep the cost of the early stages as low as possible. A large organization may choose to assign several senior people to this task, but it can also be done at a much lower level of effort. Early tracking can often be performed on a part-time basis (which offers the important ancillary benefit that the technical person so assigned generally views the opportunity to learn as a real bonus). If interest becomes serious enough, it can be pursued with the full-time assignment of a staff person. Even a pilot application can often be developed on a very limited budget. Concern about the high cost therefore need not inhibit an organization from engaging in limited technology assessment.

Early development work generally requires support from the central MIS group to spread the risk and encourage users to participate in pilot studies. Support usually takes the form of assistance from the central technical staff, but may also include funding for such things as an outside consultant or purchased hardware and software. This funding helps subsidize the cost of organizational learning and the diffusion of new technology to other parts of the firm. Organizational learning, rather than a working application, is often the primary justification for a pilot project, and so costs of the learning should be shared across the organization as a whole. The central staff can further accelerate learning and technological diffusion by participating in the careful assessment of each pilot project—its costs, benefits, and problems of introducing it.

Central support should be limited to a few of the most promising technologies rather than spread across a broad front. One should choose pilot projects that offer the most favorable set of characteristics, such as the availability of the necessary technical skills, strong user support, and the prospect that the lessons learned will find broad applicability across the firm.

It is only after a technology has been well assimilated through one or more pilot projects that the firm should consider a large investment in a full-scale production version. Learning and mistakes should come when they are relatively cheap, not in a large project where the stakes grow enormously. Success in technology deployment depends critically on this obvious but often ignored concept.

Further Readings

Galbraith, Jay, *Designing Complex Organizations,* Addision-Wesley, 1973. One of the few books on organizational design that recognizes the critical importance of the organization's information-handling capabilities.

Head, Robert V., *Strategic Planning for Information Systems,* Q.E.D. Information Sciences, Wellesley, MA, 1982. A sound discussion of MIS planning, written by a respected senior authority on MIS.

McLean, Ephraim R. and John V. Soden, *Strategic Planning for MIS,* John Wiley & Sons, 1977. A discussion of planning in general and MIS planning in particular, with numerous examples drawn from the public and private sectors.

Markus, Lynne M., *Systems in Organizations: Bugs and Features,* Pitman, 1984. An interesting discussion of the behavioral issues in implementing information systems.

Sprague, Ralph H. Jr. and Barbara C. McNurlin, *Information Systems Management in Practice,* Prentice-Hall, 1986. A comprehensive and very practical treatment of many of the important issues in organizing and running an MIS in organizations.

Van Schaik, Edward A., *A Management System for the Information Business: Organization Analysis,* Prentice-Hall, 1985. Management of the MIS, with special relevance for larger organizations.

11

Implementing A Successful MIS Strategy

The idea of gaining a competitive advantage through the application of information technology has become thoroughly fashionable. A few years ago the notion would probably have struck most senior managers as absurd; now many of them devote considerable resources and energy to making it happen. The potential rewards certainly justify this attention, but effort alone offers no guarantee of success.

A number of powerful concepts, insights, and analytical methodologies have emerged from attempts to better understand the role of information systems in organizations. It is quite likely that many of the general ideas will survive as we acquire a more comprehensive understanding of information systems. In this last chapter of the book it is appropriate to examine some of the elements that appear to be key to a successful MIS strategy.

Critical Success Factors

The Concept

A significant competitive advantage can only come from activities that contribute to an organization's basic business functions. This simple idea nevertheless took a long time to have much effect on the way most organizations allocate their MIS resources. In fact, the bulk of their MIS activities still deal with back-office functions having little to do with strategic goals.

Establishing a link between fundamental business needs and the MIS requires an explicit statement of those needs. A methodology that has proved very useful is to have management define the organization's **critical success factors**, or **CSFs**. These are the relatively few things that an organization judges that it must do well to thrive. Low prices, high quality, differentiated products or services, advantageous sources of supply, fast response to customer orders, marketing effectiveness, quality of human resources—all are candidates for adoption as CSFs by various firms (but not all by the same firm, since they conflict in significant ways). Even successful organizations in the same industry might have quite different CSFs, because there can be many roads to success. The choice of CSFs is more a policy issue than one intrinsic to an industry.

The Methodology

John Rockart and his colleagues at M.I.T. have developed and successfully applied an extensive methodology for eliciting CSFs from management. Using semi-structured interviews with managers throughout the organization, analysts develop a hierarchy of self-consistent CSFs appropriate for each organizational unit. To concentrate a unit's efforts on the really important matters, each manager should generally limit the number of CSFs to half a dozen or so.

The CSFs serve as the basis for defining critical information needs. Suppose, for example, that product quality is defined as a CSF for a manufacturing company. This triggers a detailed analysis of how the information system can contribute to improved quality. It can:

- Maintain detailed records of the quality of material received from each supplier as a basis for supplier selection and negotiation.

- Capture inspection report data at each stage of the production process with the view of identifying and correcting sources of quality problems (e.g., a defective product design or a faulty manufacturing process).
- Analyze defective items returned under the warranty program to improve product design and manufacturing techniques.
- Analyze replacement part orders to identify usage patterns that indicate abnormal wear or breakdown.

Information technology cannot necessarily make a significant contribution to every CSF, but it does in a surprising number of cases. A college that regards greater alumni contributions as a CSF, for example, might explore how a sophisticated database management system could personalize alumni relations by maintaining data about an alumnus' past associations with the institution (e.g., as a student, past giver, parent), along with other background information (e.g., company affiliations, personal honors, publications). Some organizations have found that brainstorming sessions among involved parties can often identify a number of imaginative ways in which the technology can help.

An important aspect of the methodology is choosing one or more quantitative measures for each CSF. For example, a composite index of quality might be used for the product quality CSF, with weighting for such factors as the severity and recency of defects. The necessary raw data should generally be captured within the operational system, and so the analysis and reporting of CSF accomplishments present no great technical or economic difficulties. By far the more difficult part is getting management to come to grips with its part of the process—formalization of the critical determinants of success and the quantification of suitable measures of performance.

Competitive Strategies

Try as they may to escape it, virtually all organizations confront a competitive market. This comes as no surprise to the typical firm in the private sector, but it is also being increasingly realized in many nonprofit organizations that may have traditionally viewed themselves as above the fray of the market. Hospital and college administrators, for example, certainly recognize that they face, at best, a genteel competition for clients and funds. Even organizations formerly immune from direct competition, such as regulated monopolies, risk growing competition from substitute products or new entrants in increasingly unregulated markets.

Elements of the Competitive Market

Even though different organizations may choose quite different ways to compete, the motivation is much the same. Michael Porter's seminal work on competitive strategy provides a sound starting point in understanding how information systems can contribute to an organization's competitive position.

In brief (with due apologies for oversimplification), Porter identifies the following characteristics that govern an organization's competitive position:

1. Efficiency and effectiveness of the organization's internal operations.
2. Relationship of the firm to its suppliers (in terms of bargaining power, degree of coordination, etc.).
3. Relationship of the firm to its customers (e.g., bargaining power, degree of coordination, cost of switching to a competitor, etc.).
4. Exposure of the firm to the entrance of new competitors (e.g., a national retailer entering the insurance business).
5. Exposure of the firm to substitute products or services (e.g., a bank subject to the competition of a brokerage firm that offers a set of integrated financial services).

Within this competitive structure, a firm can choose to compete through some combination of price and product/service differentiation. In the case of a product, it is impossible to separate its physical properties from such service characteristics as the speed and reliability of delivery, credit terms, and return privileges; the customer sees them as a bundle. A retailer, for example, may base a successful business on selling commodity products supported with differentiated services.

Use of Information Technology for Enhanced Competitiveness

Porter's framework provides a useful basis for focusing attention on fruitful areas for improvement. In pursuing a strategy of price competition, for example, a manufacturer might direct efforts at reducing supplier costs, increasing internal efficiencies, or entering long-term fixed-price contracts with customers. Information technology has obvious relevance in formulating and implementing such a strategy. The following illustrate just a few of the possible ways that the MIS could contribute:

- Use of telecommunication links with major suppliers to provide economies of a more stable production level and reduced buffer inventories.
- Analysis of supplier prices, delivery costs, quality, reliability, and credit terms to lower overall material costs.
- More efficient production scheduling to lower the costs of direct labor and increase effective capacity.
- Use of on-line order entry from the field (possibly by the customer personnel directly from their own premises) to reduce processing lags, thus allowing fewer regional warehouses while maintaining or improving average delivery times.
- Analysis of data derived from the operational system to develop tactical and strategic plans dealing with such matters as plant locations, capacity planning, make-or-buy decisions, and the cost savings from long-term sales contracts that stabilize production.
- Sophisticated analyses of a product's cost structure and price elasticity to recommend more effective pricing strategies.

Information technology also contributes to a strategy of product/service differentiation. Unlike standardization, which aims at conserving information processing, differentiation requires the use of information to tailor a product or service to the idiosyncratic characteristics of a specific customer or limited market. A made-to-order suit, for example, requires much more information processing—in the form of individual measurements and fittings—than an off-the-rack suit.

Numerous examples exist of information-supported differentiation:

- An order entry system for scheduling an automobile for specific customers wanting specific sets of options.
- A CAD/CAM system that lowers the cost of customizing products to the point that they can be offered at prices competitive with standard products.
- A life insurance company, in countering competition from other types of financial service companies, develops products that combine term life and mutual funds based on a sophisticated computer analysis of a client's detailed financial status.
- As part of its credit card system, a bank develops a detailed database about individual customer credit and payment patterns, enabling it to incorporate humanlike intelligence in making judgments about such matters as accepting a credit charge or sending a dunning letter to a tardy payer.
- A mail order company in the gourmet food business maintains a detailed database of its customers' past purchases to generate tailored direct mail announcements and offer individualized services (such as a dessert-of-the-month club customized to a customer's preferences).
- On-line telephone order taking for the gourmet food company allows the salesperson to give an immediate response concerning stock availability and shipping dates, provide suggestions for complementary products (cornichons to accompany a hearty pâté, say), and evince personalized interest in the customer (e.g., asking whether he was happy with the latest dessert-of-the-month delivery).
- Rapid order processing and warehouse picking to expedite delivery to customers (with an option offered to a customer of paying a premium for express service).
- An airline reservation system allowing customers to request special meals to meet individual dietary requirements (with a provision for frequent flyers to have standing requests for special meals, fare class, choice of seat location, credit card billing, etc.).

- Add-on services derived as a by-product from a basic customer service, such as a telephone billing system that gives a customer the option of receiving her bill summarized in user-specified ways—for example, by area code, calling period, duration of call, or cost of call.

A strategy of competing through product and service differentiation offers the vastly appealing prospect of allowing a firm to escape from low-margin price competition. It can be built on sophisticated and evolving technology that many competitors cannot match. When seeking a sustainable competitive advantage from its MIS, a company would do well to explore this avenue.

The Value Chain

It is difficult to identify the most attractive areas in which to pursue a competitive advantage through information technology. Valuable insights can be gained by examining a product's **value chain**—the stages by which value accumulates on the way from suppliers through to the final customer. This analysis identifies where value is added in the process, thereby allowing management to concentrate attention on the high-value areas and to consider possible ways to gain a competitive advantage by shifting functions to different parts of the value chain.

Figure 11.1, based again on Porter's work, shows how value typically gets added in a manufacturing firm. It includes the following steps:

1. The inbound logistics: movement of raw materials from the supplier to the firm's internal operations.
2. Internal operations: conversion of raw materials into finished products.
3. Outbound logistics: distribution of finished products to customers.
4. Marketing and sales: customer relations, processing customer orders, etc.

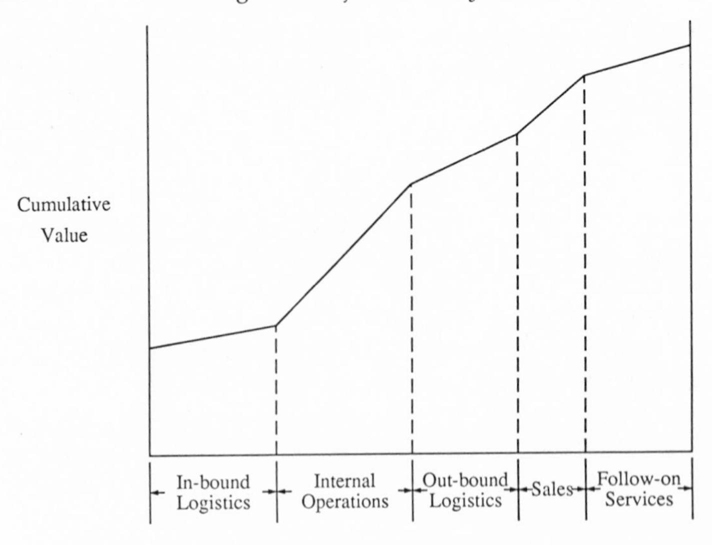

Figure 11-1. Value chain.

5. Follow-on services: maintenance, product support, product disposal or recycling, etc.

A value chain analysis reveals where substantial leverage can be achieved. A company producing a bulk chemical commodity, for example, found that outbound truck deliveries accounted for about 15% of the total value added. Determining a delivery schedule involved many producing sites, many customers, high storage costs, and complex constraints on the timing of deliveries. Given the complexity of the task, human dispatchers were able to do an astonishingly good job. A mathematical model, however, improved the plan by as much as 10% by considering many more variables and alternatives than a human dispatcher could. A 10% savings in a 15% value-adding function—that is, a saving of 1.5% of the total—starts to get management's attention. If the firm started with a beginning profit margin of 10%, a 1.5 saving on gross sales amounts to an increase in net profit of over 13%. That is a fairly strategic gain by most reckoning, especially in view of the fact that the saving was accompanied by improvements in customer service that further enhanced the company's competitive position.

Providing Customers with Added Value

The value chain between a firm and its customers deserves special attention, because it offers numerous opportunities for gaining a competitive advantage. Furthermore, an understanding of this portion of the value chain also sheds light on supplier relationships, for one is the mirror image of the other (with the role of the firm reversed from seller to buyer).

The information system can be instrumental in shifting value-adding functions from buyer to seller, to the benefit of both parties. Consider the following examples:

- By maintaining a detailed database about past sales of each customer, a magazine distributor is able to generate recommended order quantities for each newsstand.
- A distributor of products to drugstores uses past sales data to make ordering recommendations to each customer, and delivers the products with pre-attached price labels (based on customer-specified profit margins for each product category) in the store's stocking location sequence (to reduce the labor in stocking the shelves).
- A heating oil distributor maintains a record of each customer's consumption pattern, the date of the last refill of the storage tank, and the weather conditions since the last refill; based on these data, the distributor then chooses when to make the next refill delivery.
- An airline supplies accounting services to its travel agent customers, employing the existing terminals and communication lines normally used to make passenger reservations.
- A parts supplier maintains sufficient inventory to allow an assembly plant to significantly reduce its buffer stock (in a "just-in-time" inventory system).

The shift of a function can go the other way as well—from seller to buyer. The classic case is the telephone company's use of automatic dialing, with the subscriber taking on the work formerly done by telephone operators. The modern version of this is the remote interactive terminal installed in a customer's premises, with the customer performing the data entry function and bearing responsibility for its accuracy (but gaining the compensating benefits of immediate feedback on prices and delivery).

These examples have an important characteristic in common. They all represent joint efforts between seller and buyer to achieve a more efficient or effective allocation of functions. A shift of functions from one party to the other might be warranted for one of the following reasons:

1. One party has easier access than the other to data, computational facilities, or skills (e.g., the magazine distributor has easier access to sales data and the facilities to analyze them than does the newsstand proprietor).
2. One party is in a better position than the other to exploit economies of scale (e.g., the drug distributor can use automated price labeling equipment and the airline can eliminate duplicate accounting systems among travel agents by developing a common system).
3. One of the parties can eliminate a function previously performed by the other (e.g., the drug distributor can load trucks in shelf stocking sequence to eliminate rearranging stock by the store—requiring, to be sure, a sophisticated warehouse system in order not to merely shift costs from the retailers to the distributor).
4. Closer coordination between the parties can reduce combined buffer stocks and slack resources (e.g., stocking of inventory by the supplier).
5. A function can be moved to the party that has the most control over its quality (e.g., data entry from customer terminals).

The fact that all these examples depend utterly on sophisticated information processing is perhaps the most significant common characteristic of all. Information technology—especially communications technology—breaks down rigid corporate boundaries and permits a shift of functions to the part of the joint system that can best perform them.

These shifts raise the issue of how to share the benefits coming from joint actions. Often the supplier of a new service simply charges a fee. In other cases the supplier may not charge an explicit fee, but uses the service as a means to gain a competitive advantage through differentiation (which is all the more sustainable if the service is derived from a continuing relationship that competitors cannot share, as in the example of the drug distributor with unique access to knowledge about its customer's business). If a function is shifted to the customer, compensation may take the form of a lower price or improved service (e.g., a dialed telephone call is cheaper and faster than an operator-assisted one).

Identifying Strategic Opportunities

Frameworks such as CSF and value chain analysis provide a valuable structure for problem solving, but certainly do not automatically generate imaginative ideas about how the information system can strengthen the strategic posture of the organization. Such ideas can only come from creative work by managers and workers throughout the organization. How can this process be made more productive?

In making suggestions for MIS applications, users need to shed some of the misconceptions they often hold about computer systems. We have conditioned them through unhappy experience to believe that information processing is expensive, inflexible, unruly, and pedestrian. In fact, with the right implementation, just the opposite statement gives a truer picture:

- Because of the low cost of information processing, the organization should try where possible to substitute information processing for other resources (e.g., people, inventory, plant capacity, real estate).
- A well-designed MIS provides a far greater capacity for learning and adaptation than a nonautomated system.

- With end-user programming, users can satisfy many of their own information needs.
- An MIS is relevant to some of the most exciting and critical functions of the business.

None of this is at all obvious to most managers to whom we must look for suggestions about fruitful MIS applications. It is therefore critical that they gain more contemporary insights about information systems. Favorable direct experience with a well-managed MIS clearly offers the most convincing source of these insights, but a formal program of user education is an essential first step.

Also needed are formal consultative mechanisms for getting user inputs. A **steering committee** composed of representatives from major constituencies provides a common advisory forum. Such groups are needed at all levels in the organization and throughout all stages of the implementation and operation of a system. At the highest level, a committee is needed to set basic policies, priorities, and funding limits. At the lower levels, the committees deal with detailed functional specifications and coordination across organizational units. Most organizations have found it valuable to establish a regular meeting schedule with formal agendas.

Managers throughout the organization should be called on to identify explicit links between their business plans and the MIS. If indeed the MIS is to play a central role, it should be able to contribute widely to furthering those objectives that managers establish as most critical. A business plan without identified MIS support may very well be overlooking important opportunities. The issue of such support should at least be addressed.

In trying to get users to strike out in bold new directions, we need to apply mechanisms that both engage their interest and liberate their thinking. Brainstorming, game playing, role playing, and structured problem solving methodologies have all been used to good effect in eliciting creative ideas. One might ask a manager, for example, how she would run her organization differently if computer processing were entirely free (which is not too far off the mark compared to past technology and current perceptions).

Allocating Resources to Information Processing

Routine transaction processing accounts for the lion's share of most current MIS budgets. Because of its high volume and growing penetration into all operational activities, transaction processing will continue to demand an important fraction of information processing resources. Several contemporary developments, however, substantially alter the traditional allocation.

Some of the new functions added to MIS responsibilities put major new demands on resources. The cost of voice communications alone generally swamps the entire budget for data processing. The office information system, when it places a workstation on almost everyone's desk and links it to a distributed network, can cost as much as 10% of the firm's salary budget—again, quite possibly exceeding traditional data processing costs. New initiatives to develop decision support systems (including sophisticated expert systems) add still further resource demands.

Not all information processing costs show up on the MIS budget, of course. An increasing fraction, in fact, will doubtlessly come out of user budgets to pay for personal workstations, shared departmental machines, and local technical support (not to mention the hidden cost of users' time). Regardless of how these costs are budgeted, information processing in all its forms will claim a growing share of the organization's resources. The issue clearly merits top-management attention.

Most MIS activities should continue to be funded through the organization's normal resource allocation procedures. Major investments in information systems should be subjected to the type of cost-benefit analysis discussed earlier, with the beneficiaries of a system generally paying the costs of its implementation and operation. Decentralization of decision making is based on the assumption that lower-level managers have sufficient incentives to husband their resources, and there seems little reason to alter that basic principle in the special case of information processing.

There are, however, good reasons to provide top-level review and limited financial support for information processing activities. We have already discussed the need for central funding of high-risk projects having potentially broad application within the firm. The amount of the funding required for this purpose is generally quite modest, but it can provide significant leverage in accelerating strategic innovations. The top-level MIS steering committee should give serious attention to the strategic and policy implications of centrally funded projects to determine that long-term needs are being met. Periodic reports on these matters should be given to the firm's top executive councils.

Senior management should ascertain whether adequate funding is provided to develop the telecommunications network and data sharing mechanisms. This core infrastructure provides enabling capabilities that facilitate all other developments. Because the cost of developing the infrastructure is quite high, often without much immediate payoff, corporate funds must generally be allocated to the task.

Some attention should probably be given to MIS expenditures on the part of decentralized units (although one should certainly shrink from imposing unnecessary red tape or reporting requirements). Senior management should stay informed about the general allocation of information processing resources. Expenditures might be classified and reported along several different dimensions:

- By organizational unit.
- By critical success factor.
- By function: core infrastructure, transaction processing, decision support, office automation.
- By motivating purpose: meet basic needs of business, satisfy imposed requirements (legal requirements, say), counter a competitive threat, gain a strategic advantage, build new skills.
- By competitive strategy: lower costs or provide differentiated products or services.

- By point in the value chain: inbound logistics, internal operations, outbound logistics, sales, and follow-on services.

Other dimensions might also be devised to provide additional insights into MIS activities. Certainly only a few such classifications are likely to prove useful enough to justify the expense of collecting and reporting the information. The boundaries between different categories of expenditure are generally quite fuzzy—and becoming more so with increased integration of the MIS—so some arbitrary allocations will inevitably be required (but ideally with sufficient consistency to spot trends over time).

Managing Through the MIS

Most executives find it difficult to view the information system as an integral part of their decision and management process. Acquiring information and effecting action through a comprehensive MIS, rather than through people, call for a systematic and abstract approach to problem solving that most of us find unfamiliar and uncomfortable. And yet, the ability of an organization to leverage its human resources through information technology depends heavily on widely instilling the concept of the MIS as the core management mechanism of the organization—its central nervous system, as it were.

Suppose, for example, that management suddenly judges that inventory levels are dangerously excessive. One can—and probably should—take immediate direct action to deal with the problem by cutting back on production and purchases. If those were the only steps taken, however, the organization commits itself to similar firefighting crises in the future. A far more fundamental approach, after the immediate crisis has passed, would be to change the information system in ways that reduce the probability and severity of future problems. A number of ideas come to mind:

1. Develop an exception reporting system that sends earlier signals of impending problems and identifies their source (e.g., sales lower than expected or production higher than planned).
2. Improve the forecasting system.
3. Speed up the logistics system so that production schedules can be modified quickly to respond to forecast errors.
4. Build in a mechanism that allows management to choose an aggregate inventory level and have the system translate this decision into the detailed actions necessary to achieve the specified total figure.

Figure 11-2 depicts this central role of the MIS. Events in the real world—arrival of customer orders, receipts of material, completion of manufacturing operations, and the like—are

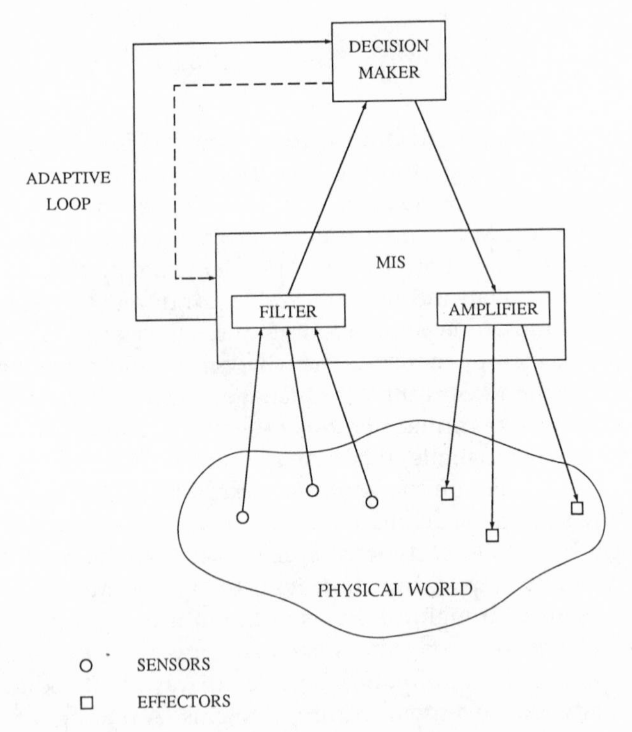

Figure 11-2. An idealized management system.

sensed through the operational part of the MIS. The resulting detailed data are then used in processing the various transactions at the operational level, and the appropriate actions are taken (shipping to customers, paying suppliers, etc.).

The MIS serves as a filter between the raw operational data and human consumers of information. Many operational matters may be handled with little direct human intervention—for example, through robots in the factory and automatic distribution over a network (such as a purchase order transmitted directly to a supplier). Information displayed for a human decision maker will come from a filtering process that transforms a vast database into highly selected information (using summary reports, ad hoc queries, exception reporting, and graphic representations).

The resulting decisions get executed by passing them down to lower levels in the hierarchical system. Here they are translated into greater detail, based on prescribed models and rules. The high-level decision about aggregate inventory, for example, could be translated into item-by-item order points and order quantities that in total are expected to conform closely with the specified total amount of inventory. The bottom level in this hierarchical process actually executes the plan through various effector mechanisms (e.g., by controlling valves in the plant, issuing purchase orders with the revised order points, displaying operating instructions to human operators, etc.) The system automatically monitors adherence to current plans, and reports significant deviations as part of the filtering process.

A complex system of this sort can never work perfectly, certainly not in its early versions. It can get better only through a long period of learning and adaptation. There is a critical need, therefore, for a design that facilitates such adaptation. This places a high premium on the use of techniques that increase the system's flexibility, such as 4GLs and prototype implementations that can be readily scrapped and rebuilt better the next time.

Decision makers contribute to the adaptive process by controlling the filters and amplifiers built into the system. They generally exercise control through "dials"—for example, parameters selectable from a menu display on their workstations—that permit adjustments to the filtering and amplification rules. More basic adaptation comes through modifications in the programs that perform the filtering and amplification.

An important aid to effective adaptation is the self-diagnosis that the system provides about its own operation. It should generate detailed statistics about such matters as the volume of each type of transaction, volume of telecommunications traffic analyzed by sources and destinations, capacity utilization, detected data entry errors, detected violations of security, hardware or software failures, response times, and chargeback costs. Designers can then use this information in a process of continual refinement.

A Parting Word

The need was stressed earlier for a guiding vision to motivate and sustain the long-term effort necessary for an organization to gain a strategic advantage through its MIS. The characteristics of a visionary system were outlined to suggest the general nature of future systems as they evolve from their present roots.

The natural tendency might be for management to dismiss such a vision as too ambitious, too distant, or simply impractical. That would be, I believe, a serious mistake. Any organization can at least begin some early actions to take better advantage of emerging opportunities. These actions could initially be kept fairly modest in scope, consistent with the needs and resources of the organization and the confidence that management has in its ability to point in the right direction.

The willingness of management to set out on a long-term program to exploit information technology depends critically on its expectations about the importance of the MIS in contributing to the organization's basic goals. Management can indeed have great confidence about the following expectations:

- Information technology will continue to advance at a dramatic rate similar to the last four decades.

- Information systems will gain an accelerating importance in all aspects of managing an enterprise.
- An organization that masters the process of installing a successful information system is likely to gain significant strategic advantages—or avoid a significant disadvantage.
- No one can predict the precise information needs of an organization, and no one can build a system right the first time; therefore, a good system can evolve only over a long period of learning and adaptation.
- Successful systems can only be built on a foundation of a sound infrastructure that enables the organization to implement cost-effective applications and adapt them as needs and perceptions change.

There is not much uncertainty or ambiguity about the early steps that management should take if these assertions prove even partially correct. If an organization has not already done so, it should begin to develop the human resources and technical infrastructure that enable it to build a strategic information system. It should start to create a shared vision among its stakeholders of how it might deploy information technology to its greatest advantage. It should set up mechanisms for learning and adaptation to position the information system to meet evolving needs.

It should begin with deliberate dispatch the first step of a long process.

Further Readings

McFarlan, F. Warren and James L. McKenney, *Corporate Information Systems Management: The Issues Facing Senior Management*, Richard D. Irwin, 1983. Sound treatment of how the MIS can provide competitive advantages.

Porter, Michael E., *Competitive Strategy*, Free Press, 1980. This book, along with Porter's 1985 book listed below, have had a major impact on how organizations view their competitive strategies. Although by no means confined to MIS issues, the Porter framework has been widely discussed in connection with gaining a competitive advantage through information technology.

Porter, Michael E., *Competitive Advantage,* Free Press, 1985.

Rockart, John F. and Christine V. Bullen (eds.), *The Rise of Managerial Computing,* Dow Jones-Irwin, 1986. Contains several papers on the Critical Success Factor methodology, with case examples.

Synnott, William R., *The Information Weapon: Winning Customers and Markets with Technology,* Wiley, 1987. A management-oriented discussion of the strategic use of information systems, written by an experienced practitioner.

❖ ❖

Glossary

The numbers in parentheses at the end of each entry give the pages in the text where the glossary term appears in boldface.

4GL: *See* Fourth-generation language.

Access time: The time required to begin transferring data from an external storage medium into the main memory of a computer. (36)

Accuracy: The degree to which a numerical value matches the actual physical quantity it purports to measure. (209)

Active data dictionary: A data dictionary linked directly to a database management system such that any change in the dictionary automatically governs access to the database. (188)

Ad hoc query/report. A request for information not anticipated in advance and "wired in" as a standard report. (105, 116)

Address: A numeric value that provides a unique identifier of the physical location of a character or word stored in the computer's main memory; also the location of a sector or record stored in a direct access storage device. (65, 97)

AI: *See* Artificial intelligence.

Algorithm: A computational procedure for performing a specified task, such as sorting a set of names into alphabetical sequence. *(72)*

Analog: A physical quantity (e.g., voltage or length) used to represent a numeric value. (6)

Analyst: A person responsible for translating functional needs of an information system into a set of detailed program specifications. (68)

Application generator: A high-level computer language used to produce application software; generally considered an example of a "fourth-generation" language. (73)

Application program: A computer program that performs the tasks associated with a given end-user function. (26, 67)

Architecture: The basic form and relationships among the components of a computer system, analogous to the form and relationships among the components of a building. (76)

Archival storage: Storage of seldom-used records in an inexpensive but relatively inaccessible storage medium, such as magnetic tape or microfilm; often used to retain inactive historical data or to satisfy legal requirements for keeping data for a specified time period. (39)

Artificial intelligence: The branch of computer science that deals with the use of a computer to perform humanlike functions, such as playing chess, diagnosing patients, or setting premium rates on casualty insurance. *See also Expert system.* (119)

Assembler: A computer language translator that converts a source program written in a symbolic machine-oriented language into an object program in the machine language of the computer on which the program is to be run. (73)

Auxiliary storage: *See* External storage.

Backup: Duplicate data or hardware used to replace the primary resources in case they are destroyed or temporarily out of service through error or system failure. (190)

Bandwidth: The information carrying capacity of a communications channel, measured in bits per second. (The term *baud*, generally treated as synonymous to bits per second, is also used as a measure of bandwidth). (49)

Batch: A set of transactions that are collected over a period of time (such as a day) and then processed together. (95)

Batch processing: The processing of transactions in batches, often on a scheduled periodic basis (e.g., daily or monthly). (31)

Benchmark program: A program executed to calibrate the capacity and response time of a computer system (to compare one vendor's system with another, for example). (164)

Binary: A method of representing characters or numeric values using a two-state (0/1) coding scheme. (5)

Bit: A single binary digit (0 or 1); a measure of information capacity. (35)

Bit-mapped: A method of translating an image represented in binary form within the computer's memory onto a display screen, in which each bit in memory corresponds to a specific dot or *pixel* on the screen's surface. (36)

Boundary: The separation of functions between a system (or subsystem) and its environment. (241)

Branch instruction: A computer instruction that allows the computer to alter the sequence in which instructions are executed, generally based on a test for certain values of input data (a stock balance of zero, say). (66)

Buffer: Temporary storage for cushioning the effects of *coupling* between subsystems. (248)

Bug: An error in the logic of a computer program that results in a malfunction or erroneous result when the program is executed. (149)

Bundled pricing: An approach to pricing in which all products associated with a computer purchase are included in a lump-sum price. (68)

Byte: A contiguous collection of eight bits used in the binary representation of numeric values or coded data; often used synonymously with the term *character,* since one byte is usually used to represent one data character (out of a character set of 256 possible values). (35)

CAD/CAM: Computer-Assisted Design/Computer-Assisted Manufacturing; a computer system used in the design of a product and its subsequent manufacture. (10, 277)

Central processing unit: The part of a computer in which instructions are interpreted and executed. (41, 63)

Chargeback: A cost allocation method by which users are charged for the computing services they receive. (233)

Chief information officer: The executive having corporate-wide line and/or staff responsibility for such information-intensive activities as data processing, voice and data communications, office information systems, and external information services. (23, 279)

Chief programmer: The person in charge of a team of computer programmers, who is responsible for the overall design of a program and the supervision of other team members. (157)

CIO: *See* Chief information officer.

Clone: *See* Plug-compatible product.

COBOL: COmmon Business-Oriented Language. The most widely used third-generation procedural language for programming data processing applications. (68)

Coding: Representing the logic of a computer-based algorithm or function in the precise form and syntax used for a given computer language; programming. (148)

Command language: A form of dialogue with the computer in which the human inputs commands for execution by the computer (as opposed, say, to selecting an option from a menu of alternatives presented by the computer). (117)

Compiler: A computer program that translates a source program written in a higher-level language (e.g., Fortran) into a machine language object program for the machine on which the program is to be executed. (73)

Computer: An electronic device for automatically executing a computational algorithm defined by a series of instructions stored in the computer's memory. (5)

Computer language: *See* Programming language.

Connectivity: The ability to establish telecommunication links among a variety of equipment—computers, terminals, storage devices, and the like—in order to exchange data and share resources. (281)

Constraint: A maximum or minimum value that a variable is permitted to take, either because of physical limits on a resource or because of policy limitations. (108)

Control system: A system that monitors the performance of specified variables in a process and then calls for appropriate corrective action if a variable is found to be outside of its allowed limits. (243)

Coordination: Communication among subunits of a system in order to choose actions within each subunit that are consistent with the overall goals of the system. *(250)*

Coupling: The linking of subunits within a system that occurs when an output of one subunit serves as an input to another subunit. (243)

CPM: Critical Path Method; a technique for scheduling and controlling a complex project consisting of numerous interconnected tasks in which some specified task must be completed before other tasks can begin. (151)

CPU: *See* Central processing unit.

Creeping commitment: The step-by-step implementation of a project, with each step preceded by an evaluation of whether the overall project should be continued; if the decision is made to continue, a budget, schedule, and required outputs are set for the next step. (231)

Critical success factor: An activity that the organization must perform well to meet its long-term goals. (9, 290)

CRT: Cathode Ray Tube; a terminal device for displaying data in transient form on a TV-like screen. (27)

CSF: *See* Critical success factor.

Cursor: A rectangle or similar mark shown on the surface of a display screen to identify the point on the screen at which editing functions take place (such as typing in new characters or deleting existing characters). (181)

Data: Numeric values, textual characters, and graphic images, generally retained in the database in raw form prior to processing into information for operational or decision making purposes. (106)

Data dictionary: A computer-based depository of information that describes the content, meaning, format, logical structure, and security protection of the data stored in a database. (187)

Data element: An individual piece of data, such as an employee's name or a customer's account balance; also called a data *field*. (91)

Data entry: The function that deals with the input and editing of data entering the MIS. (27)

Data hub: A central depository of shared data. (51)

Data independence: The degree to which data in a database can be changed without having to make corresponding changes in application programs (and vice versa). (189)

Database: The collection of machine-readable data maintained as part of a management information system; the collection of data described by the system's data dictionary. (12, 24)

Database administrator: A person assigned responsibility for managing the organization's data resources. (188)

Database management system: A system software product that provides a variety of functions needed to access, maintain, and protect the organization's database. (18, 67, 183)

Data-oriented DSS: A decision support system designed to give a decision maker selected access to the database. (105)

DBMS: *See* Database management system.

Debugger: A software product used by a programmer to test a program, analyze the code to identify bugs, and assist in making program changes. (160)

Decision support system: A computer-based system designed to assist a decision maker (or group of decision makers) to make better, faster, or cheaper decisions. (86, 100)

Decoupling: A means of providing partial separation between subunits of a system that reduces the degree of interaction due to coupling. (248)

Default: A value used automatically by a system in the absence of an explicit choice by the user. (175)

Deployment: The process of converting a new application program to operational condition. (150)

Deterministic value: A value known with certainty; a single fixed value used as a simplifying approximation for a variable subject to some probabilistic variation. (132)

Dial-up: A one-time communication link established (generally for a relatively short duration) through a dialing and switching process (as in a standard telephone call). (50)

Digital: The representation of data in digital (i.e., numerical) form, generally using some form of binary coding. *See also* Analog coding. (5)

Direct access processing: Processing of data stored on a direct access storage device, allowing transactions to be processed individually rather than in a batch. (94)

Direct access storage: An external data storage medium—most commonly magnetic disk—that allows the computer to store or retrieve a record without having to access all intervening records (usually within a small fraction of a second). (38)

Distributed system: A system composed of multiple computers connected through a telecommunications network. (48)

Document processing: Activities associated with the electronic creation, storage, retrieval, and distribution of textual or image documents. (277)

Documentation: Written text to describe the purpose, functions, logic, and operation of an information system and its constituent programs. (148)

Dot matrix printer: A printer that forms the shape of each character from a matrix of dots printed by small pins striking the paper. (43)

DSS: *See* Decision support system.

Econometric database: A database of economic variables, such as a country's gross national product over a period of time, broken down by industry sector. (105)

Economic order quantity: The replenishment quantity of an inventory item that results in the lowest overall cost. (122)

Editing: The process of verifying the accuracy of input data and taking appropriate corrective action when errors are detected; the process of creating or changing textual data on a computer. (29)

Effectiveness: The degree to which an action leads to a desired end result. (213)

Efficiency: The level of resources used to accomplish a given end result, compared to the minimum possible level that could have been used to accomplish the same result. (213)

Efficiency frontier: The collection of system designs that provides the most efficient means of satisfying a range of specifications. (210)

Electronic mail: The transmission of text (or possibly images) in interpersonal communication over a telecommunications network, allowing either a simultaneous conversation or (more typically) message storage with delayed retrieval on demand. (139)

Electronic publishing: Preparation in hard copy form of textual or image information stored in digital form, often operated as a central service. (*Desktop publishing*, in contrast, is usually done on the user's personal computer.) (277)

Enabling technology: A set of basic technological capabilities that allows an organization to implement relatively quickly a desired application; effective database management and communications are generally considered to be the foundations of an MIS enabling technology. (280)

End-user computing/end-user programming: The satisfaction of a user's information needs through his or her own actions, rather than through an intermediary programmer or analyst. (153, 202)

Environment: That part of the world outside of the defined boundary of a system. (241)

EOQ: *See* Economic order quantity.

Erasable: A characteristic of a storage medium that allows it to be used to store new data in place of previously recorded data. (37)

Exception report: A report that includes information only about conditions that fall outside of defined control limits. (105)

Expert system: A branch of artificial intelligence that deals with complex decision processes defined in terms of a series of rules that mimic a human expert. (108)

External storage device: A data storage device that permits the computer to read stored data into its main memory (where all computation takes place) and write the results of computations back into the device; magnetic disk and tape are currently by far the most widely used forms of external storage. (41)

Facsimile (or **Fax**): The transmission through electronic means of a copy of a document containing either text or images, in which the source document is "scanned" by a transmitting device and a copy is reproduced by a remote receiving and printing device. (277)

Feasibility study: A relatively brief study of a potential computer application to assess whether the expected costs and benefits of the application appear to justify proceeding further with its implementation. (143)

Feedback: Information derived from monitoring an operation to assess whether the operation is in control. (243)

Field: A single element of data; *data element*. (27, 91)

Fifth generation: Advanced research in computer architecture and language, aimed at superseding current (fourth-generation) technology by building increased "intelligence" into computer software. (81)

File: A collection of records, generally of a homogeneous type. (33, 94)

First-generation: The earliest generation of hardware and software, roughly from the mid-to late 1940s; characterized by hardware circuits made with vacuum tubes and programs written in machine language. (75)

Fixed cost: Costs that remain essentially constant over a relatively wide range of activity. (233)

Flexibility: The relative ease with which a program can undergo modifications and extensions to adapt it to new needs. (209)

Floppy disk: A storage medium that records data on a flexible ("floppy") disk made of coated magnetic material. The disk can be removed from the disk drive, and therefore can be used for off-line storage of modest quantities of data and as a medium for distributing data or software among machines. (47)

Form: A printed sheet with named fields of information and space provided for entering specific occurrences of a field value; a similarly formatted display on a CRT screen for capturing data in an on-line data entry system. (27)

Fortran: FORmula TRANSlator, one of the most popular third-generation computer languages, which is used primarily for programming scientific and engineering problems. (68)

Fourth-generation language (4GL): A human-oriented computer language that embodies a number of contemporary techniques for increasing the productivity of computer programmers, often by the specification of tasks in nonprocedural form. (80)

Friendly: *See* User friendly.

Full-screen editor: A software product for entering and modifying programs and data that allows the user to perform editing functions (e.g., adding, deleting, and copying) at any point on the screen specified by a cursor that can be moved around the screen by means of directional keys or a pointing device such as a mouse. (176)

Functional structure: An approach to structuring an organization in which activities are broken down into separate functions (e.g., engineering, marketing, manufacturing, etc.). See also Product structure. (247)

Gateway: A node in a telecommunications network that serves as an interface between two or more component networks having different technical characteristics. (51)

Generality: The degree to which a wide range of functions is built in to an application program; the degree to which a wide range of problems can be effectively programmed in a computer language. (209)

Generator: *See* Application generator. (73)

Gigabyte (GB): One billion (1000 million) bytes (or, more precisely, 2^{30} = 1,073,741,824 bytes). (61)

Global optimum: The solution obtained when all variables that affect the result are set at the combination of values that yield the maximum utility. (132)

Goal-seeking: An analytical methodology that seeks the value of an input variable necessary to achieve a given output result. (178)

Hard copy: Information displayed in printed form. (43)

Heuristics: A set of "common sense" decision rules that are designed to result in effective—but almost always not optimum—decisions; employed when optimization or other more formal decision methods are infeasible or impractical because of the complexity of the decision process. (132)

Hierarchical DBMS: A DBMS in which logical relationships among entities are defined in terms of a hierarchical structure. *See also* Relational DBMS. (185)

Hierarchical structure: A tree-like structure in which each node (except for the top-most one) is linked to one and only one higher-level node, and may have any number of links with lower-level component nodes. (53)

High-level programming language: A computer programming language whose structure and syntax are not tied directly to the hardware architecture of the computer on which a program will be executed. (59, 76)

Icon: A stylized representation of an entity that conveys mnemonic information, used especially in representing such information on an interactive display screen. (81, 125)

Identifier: A unique number or string of characters used to provide an unambiguous identification of an entity such as a customer or inventory item. (33, 94)

Image: Nontextual information, such as a photograph, diagram, engineering drawing, or signature. (33, 277)

Incentive system: The scheme for providing monetary or other rewards as a means of motivating people to behave in a way deemed desirable by an organization. (261)

Information: Processed data used for decision making; a representation of reality, often in condensed form, that reduces the uncertainty about the true state of nature. (106)

Information Age: The current age, in which information processing and communications are playing an increasingly important role in industry, commerce, government, military matters, and our personal lives. (3)

Information center: A location where users can go to receive technical support for end-user computing; in addition, a center generally provides personal workstations or terminals for accessing a shared mainframe computer. (205)

Information technology: The technology associated with computer hardware, computer software, and communications—for the most part, based on the microelectronic chip and the digital representation of information. (3)

Infrastructure: A set of basic capabilities that provides a developmental framework for something—e.g., the development of a national economy, an organization, or a comprehensive information system. (74, 260)

Input: A resource that enters the boundary of a system. (242)

Input/output device: A hardware device for entering input data into a computer system and/or writing output data onto some recording medium. (43)

Instruction: A component of a machine language program that defines a single computational step, which generally specifies the type of operation to perform (e.g., add, compare, or move) and the data on which the operation is to be performed. (65)

Instruction address register: A storage location within a computer's central processing unit that contains the memory address of the next instruction to be be executed. (66)

Intangible benefit: A benefit that cannot be measured in monetary terms. (224)

Integrated services digital network (ISDN): A telecommunication system that provides a variety of services (e.g., transmission of voice, data, and images over a wide spectrum of speeds) using shared digital transmission channels. (52)

Integrated system: A system having tight coupling and/or extensive resource sharing among its component parts. (26, 252)

Integrity: The degree to which data in a database are accurate and valid; the extent to which a system is invulnerable to invasion by unauthorized persons or other computers. (190)

Interactions: The effects that one subsystem has on another because of coupling or the shared use of resources. (243)

Interactive system: A system that provides a close dialogue between a human and a computer, with the human issuing commands or posing questions and the computer responding appropriately (usually within a few seconds or even a fraction of a second). *(30, 87)*

Interface: The boundary between two subsystems that interact with one another, or between a human and a computer system. (43)

Interpreter: A language translator that permits a computer to execute statements written in a higher-level programming language—in effect, giving the illusion that the computer hardware directly executes the higher-level language. (73)

ISDN: *See* Integrated services digital network.

Kilobyte (KB): One thousand bytes (or, more precisely, $2^{10} = 1024$ bytes). (63)

Knowledge engineering: The function of translating human expertise and knowledge into a set of rules that can be interpreted by an *expert system*. (125)

Knowledge worker: A worker who produces or deals with information rather than concrete products. (7)

LAN: *See* Local area network.

Language translator: A system software product that translates a source program written in a higher-level programming language into a form executable by a computer. *See also* Assembler, Compiler, and Interpreter. (68)

Letter quality: Printer quality equivalent to a typewriter's, with fully connected characters (as opposed to a matrix printer's characters which are formed from a series of disconnected dots. (43)

Life-cycle development process: The cycle through which a computer program evolves, from its early creation stages through its eventual retirement; a typical life cycle might include the stages of feasibility study, design, programming, coding, test, deployment, maintenance, and final abandonment. (143)

Linear: A functional relationship in which the effect on an output variable is strictly proportional to the change in an input variable. (132)

Lisp: A programming language for manipulating symbols and lists, widely used in sophisticated artificial intelligence applications. (124)

Local area network: A telecommunications network used to link devices (computers, printers, etc.) within a relatively small geographic area (generally within a radius of one kilometer). (51)

Loop: A set of instructions within a computer program that is executed repeatedly for a fixed number of times or until some specified condition is encountered. (65)

Machine language: The language that the hardware of a given computer can execute directly without translation. (71)

Magnetic disk: A direct access storage medium for storing files and other parts of the database, in which data are recorded magnetically on a revolving disk. (38)

Magnetic tape: An external sequential access storage medium, generally used for storing large sequential files, backup files, and archival data. (36)

Main memory: The internal storage of a computer from which programs are executed and data are retained as inputs or outputs of a computation. (41, 64)

Mainframe: A powerful computer, almost always linked to a large set of peripheral devices (disk storage, printers, etc.), and used in a multipurpose environment at the corporate or major divisional level. (48, 61)

Management information system (MIS): A computer-based information system used in the operational management and decision making of an organization. (22)

Management science: The application of scientific methods to aid in the solution of management problems, generally involving the use of a model of some sort. (129)

Marginal cost: Incremental cost; the change in cost that occurs when the output of a process is changed. (236)

Mass storage: *See* External storage.

Mathematical programming: The use of mathematical models and computational procedures for finding the optimum solution for a management problem. (108)

Matrix organization: An organizational structure that assigns management responsibilities across two or more intersecting dimensions (e.g., by function and project). (247)

Megabyte (MB): One million bytes (or, more precisely, $2^{20} = 1{,}048{,}576$ bytes). (39)

Memory: *See* Main memory.

Menu: A list of alternative actions or data values presented on a display screen to allow a user to select among the proposed options during a dialogue with the computer. (117)

Microcomputer: A small, relatively inexpensive computer, usually dedicated to use by a single user; a personal computer or workstation. (61)

Microelectronic chip: A small semiconductor device (usually about a centimeter square) that incorporates digital circuits and logic for performing a variety of electronic functions. (5)

Microelectronics: The technology dealing with the design of electronic circuits using microelectronic chips. (3)

Microsecond (μs): One millionth of a second. (5)

Millisecond (ms): One thousandth of a second. (37)

Minicomputer: A medium-sized computer, usually serving a relatively small organizational unit or dedicated to a fairly narrow specialized task. (61)

MIPS: Million instructions per second, the rate at which instructions are executed by a given computer; a common measure of a computer's approximate raw computing power. *(42)*

MIS: *See* Management information system.

MIS plan: A plan for developing and operating an organization's MIS, which includes a time-phased estimate of the functions to be provided and the resources required. (259)

Model: A mathematical or symbolic representation of a real-world environment, used to predict the consequences of alternative courses of action in order to choose the best alternative. (8, 106)

Model-oriented DSS: A decision support system that has a model as a central component. (105)

Modular structure: A system (e.g., an organization or a computer program) divided into well-defined and relatively independent parts. (78, 147)

Module: A well-defined and relatively independent component of a larger system. (53, 147)

Mouse: A pointing device that a computer user moves on a flat surface, which causes a cursor to make corresponding moves on the computer's display screen (e.g., to identify a particular point on the screen). (81)

Multiprocessing computer: A computer system that uses multiple CPUs to increase the machine's effective capacity. The CPUs share common main memory and are managed by a centralized operating system. Each CPU may work independently on an assigned task (handling a transaction, say), or (less commonly) multiple processors may work cooperatively on the same task. (71)

Multiprogramming: The interlaced execution of multiple programs on the same computer, giving each program the appearance of having control of the computer (except for reduced capacity due to the sharing of the computer's resources). (66)

Natural language processing: Computer processing of natural language statements (e.g., in English or French) in a manner that permits the computer to carry on a dialogue with a user and make correct inferences about the user's intended meaning (or ask questions to clarify the meaning). (81, 118)

Needs analysis: The assessment of the information and functional needs of an organization as a first step in defining the specifications for an information system. (144)

Net present value: The current amount of money equivalent in value to a stream of future cash flows, either positive or negative (or both), based on a specified interest (discount) rate. (65)

Network DBMS: A database management system in which logical relationships can be defined arbitrarily between any pair of entities (as opposed to a hierarchical DBMS, in which only tree-like links are permitted). (185)

Node: A point on a network connected to other nodes through network links; the location on a network of terminals or other devices. (49)

Nonprocedural: A form of task specification in which the desired end result is stated rather than the procedure for obtaining the result. (78)

Object program: The machine language program that results from the computer translation of a *source program*. (73, 147)

Objective function: A mathematical function, part of an optimizing model, that formally defines the numeric utility of the outcome resulting from a given set of values of the decision variables. (108, 130)

Office information system: An information system that provides a set of capabilities to support the functions typically found in an office environment, such as text creation, person-to-person communications, and document storage and retrieval. (138)

Off-line storage: Data storage not connected automatically to a computer; storage that requires human intervention in order for the computer to access the data. (36)

OLTP. *See* On-line transaction processing.

On-line storage: Data storage connected to a computer in a way that permits automatic access to the data without any human intervention. (39)

On-line transaction processing: Transaction processing in which the computer has rapid automatic access to the database to enable it to provide quick responses to users. (94)

Operating system: A basic system software product for automatically managing the operations of a computer, which deals with such matters as controlling access to the computer, setting priorities among multiple users, allocating computer resources, and handling input/output operations; often termed the "traffic cop" of the system. (67)

Operational level: The lower levels in an organization at which routine operational matters are handled. (12, 24)

Operations research: Essentially the same as *management science*. (129)

Optical disk: A direct access storage medium in which data are recorded on a revolving disk in a form that can be sensed by optical means (e.g., photoelectric sensing of reflected laser-generated light). The principal advantage of such storage is its ability to store a very large volume of data in dense form and at low cost; the principal disadvantages are its relatively slow access time and (in some applications) the nonerasability of the medium. (38)

Optical scanning: The use of an optical sensor to read input data. (27)

Optimizing model: A mathematical model formulated in a way that permits the computation of the best possible value of the model's objective function. (129)

Optimum: The best possible solution of a problem within the range of decision values permitted by the constraints of an optimizing model. (108)

Order entry: The application program that handles the entry of customer orders and other associated functions. (26)

Organization: An affiliated group of people and capital resources established to achieve a set of objectives. (11)

Output: The results produced by a system. (242)

Overhead: Activities, and their associated costs, not directly involved in the production of a system's output; specifically in the case of a computer's operating system, the resources used in controlling the operation of the computer. (71)

Packet-switching: A telecommunications technique in which each message is broken into relatively small "packets," which are then transmitted independently from the originating node of the network to the receiving mode. (50)

Parallel operation: An approach to application conversion in which an old application program is operated in backup fashion along with the new replacement program until the new application has been shown capable of handling the processing on its own. (152)

Parallel processing: The use of multiple computer processors, managed by a common operating system, designed to increase the effective capacity of a machine; different machines provide different degrees of integration, from *tightly coupled* for handling cooperative tasks among multiple processors, to *loosely coupled* for dealing simultaneously with independent tasks. (71)

Parameter: A variable that controls the operation of a computer program by such means as designating which program options to execute or setting the value of a numeric constant used in a repetitive series of calculations. (165)

Passive data dictionary: A data dictionary not linked directly to the operation of a database management system. *See* Active data dictionary. (188)

Personal computer: A microcomputer (or workstation) dedicated to a single user at a time. (47)

PERT: Program Evaluation and Review Technique; a project-scheduling system similar to *CPM*. (151)

Pixel: A single dot, or "picture element," located on a display screen, used to form text characters or graphics by turning the appropriate combination of pixels "on" or "off." (35)

Planning system: A combination of procedures and methodologies, composed of both human decision makers and computer-based systems, for governing the behavior of an organization at the strategic, tactical, and operational levels. (243)

Plug-compatible product: A product (sometimes called a *clone*) that emulates the operation of a widely used product produced by another vendor; often used specifically to refer to a product that mimics a product of IBM (or possibly DEC). (84)

Pointer address: A storage address contained within a data record that gives the location of an associated record. (185)

Point-of-sale (POS) terminal: A device for interactively entering sales transactions by sales personnel at a store counter. (89)

Precision: The resolution with which a numeric value is expressed; the number of significant digits used to represent a value. (209)

Present value: *See* Net present value.

Primary memory/primary storage: *See* Main memory.

Primitive language function: An elementary operation available as a built-in capability of a computer language. (175)

Probabilistic value: A value subject to random variation, possibly having a known probability distribution; a value subject to uncertainty. (132)

Procedural language: A language in which tasks are specified by means of a step-by-step (how-to-do-it) series of operations. (78)

Procedure: An algorithm defined by means of a series of steps that lead to a desired end result. (72)

Program: A set of statements or instructions in a computer language that is intended to accomplish a specified task when executed on a computer (possibly after first being translated into machine-language form); to create a program. (41, 64)

Program library: A collection of standard programs available for performing common tasks, either independently or in combination with other programs. (161)

Program package: A software product used to perform a given application function. (69)

Programmer: A person who expresses program specifications in a programming language suitable for automatic translation into an executable program; a coder. (68)

Programming: The process of expressing a task specification in a language that a computer can execute directly or indirectly (i.e., after translating the program into machine language). (147)

Programming language: A language used by a human to define computation tasks for execution on a computer. (59)

Project: A set of activities intended to lead to a well-defined end product, such as a new application program. (142)

Project structure: An organizational structure in which responsibilities and resources are assigned along a project or task dimension. *See* Functional structure. (247)

Prolog: A computer language used in programming artificial intelligence applications in which a decision process is defined in terms of a set of humanlike rules. (124)

Protocol: A set of standard conventions followed by all parties to a communication in order that each party can correctly process and interpret a message transmission. (50)

Prototype: An interim application program, generally with limited capabilities, which is implemented relatively quickly and at low cost to demonstrate a functional capability, provide an unambiguous functional specification, serve as a vehicle for organizational learning, and (possibly) evolve ultimately into a fully implemented version. (82, 198)

Query: A request for specific information expressed in a language capable of being interpreted by a computer—generally in a structured retrieval language, but possibly in a natural language if the software has a natural language processor as part of its user interface. (105)

Random processing: An application for which transactions are processed more-or-less in the order in which they arrive into the system, generally within a short response time that allows interactive dialogue with a user. (94)

Read: To sense data recorded on an external storage medium and transmit it into the computer's main memory. (36)

Read/write head: A component of a storage device (such as a magnetic disk drive) that can sense the contents of stored data and record new data on the device's storage surface. (38)

Real-time process: A physical process in which control is exercised by a computer system within the (generally short) time span needed to take corrective action and maintain stability of the process. (28)

Record: A collection of data elements pertaining to a single entity, such as an employee or customer. (33, 91)

Relational DBMS: A database management system in which a user or program views data in terms of two-dimensional tables, with rows corresponding to records and columns corresponding to fields; relations among records are established through common values maintained in the tables. (186)

Reliability: The probability that a system will function adequately during periods in which it is intended to be operational. (209)

Remote job entry: The capability of entering batch processing jobs over a communications line from a terminal physically separate from the computer (generally near a group of users); the results of the processing are then transmitted back to the remote terminal for printing and/or storage. (46)

Reprographics: The functions associated with the reproduction and distribution of hard-copy documents. (277)

Requirements analysis: *See* Needs analysis.

Response time: The time lag from the completion of a transaction input until the computer output becomes available. (209)

RJE: *See* Remote job entry.

Robustness: The extent to which a system is protected from a major failure due to user errors or component failures. (209)

Screen formatter: A software product that makes it relatively easy to program screen formats for interactive applications; often a component of a fourth-generation programming language. (81)

Seamless interface: A link between separate application programs or software products that provides such close integration and consistency that it gives the user the illusion of dealing with a single product. (175)

Secondary storage: *See* External storage.

Second-generation: The period during the 1950s and 1960s characterized by the widespread use of assembly language programs. (75)

Sector: An addressable segment of storage on a magnetic disk track. (97)

Security: Protection against accidental or intentional penetration of a computer system by an unauthorized user. (190)

Sensitivity analysis: An analysis to determine the extent to which a given variable in a decision model is affected by changes in the values of other variables. (130)

Sequential processing: Transaction processing in which transaction records and database records are stored and processed in a sequence defined by a common data element type (called the *key*), such as customer number. (94)

Service bureau: An independent firm that provides data processing services for client organizations, either through the physical delivery of input and output media or via telecommunication links. (69)

Shared resource: The use of the same resource by two or more subunits of a system, thereby causing interactions among the sharing units. (243)

Sign-off: A formal procedure for transferring management responsibility for an application from the developing organization to the using organization. (151)

Simulation: Use of the computer to mimic the operation of a physical process to analyze its behavior under alternative conditions, often with the objective of searching for a set of decision variables that lead to satisfactory or even near-optimum performance. (108)

Slack capacity: The availability of a resource in excess of expected usage to protect against a surge in demand. (249)

Software: A collective term for computer programs. (67)

Software engineering: A set of disciplined methodologies for improving the reliability and lowering the cost of developing and maintaining computer software. (143)

Source program: A computer program expressed in a language appropriate for human task definition, which must be translated into a machine-language *object program* before the computer can execute it. (73, 147)

Spreadsheet language: A programming language for defining functional relationships and data manipulation operations among variables arrayed in a two-dimensional table. (174)

Spreadsheet model: A mathematical model written in a spreadsheet language. (63, 108)

SQL (Structured Query Language): A language for expressing an ad hoc query, originally developed by IBM but now rapidly becoming a *de facto* standard used by a variety of software vendors. (117)

Stakeholder: Anyone who has a significant stake in the performance of an organization or undertaking. (109)

Stand-alone computer: A computer not connected through a telecommunication link to any other computer or terminal device. (47)

Standard: A product or method of operation that is established for widespread use within an organization (or among several organizations) to facilitate communication among subunits, reduce duplication, promote resource sharing, or provide a consistent level of performance. (120, 275)

Standard rate: The price of a resource based on an expected level of efficiency, set of environmental conditions, and volume of activity; often used in allocating the cost of information services provided to internal units of an organization. (237)

Standardization: The process of setting and administering a set of standards. (249)

Steering committee: A panel of employees who have formal responsibility for providing guidance and suggestions (or even directives) regarding such matters as the quantity and quality of services to provide, resource allocation priorities, and design specifications. (300)

Stockout: When an inventory item is temporarily out of stock. (67)

Stored program computer: A computer whose operation is controlled by a program stored in its memory (which is the basis for controlling all contemporary computers). (4, 66)

Strategic level: The highest level in an organization, which deals with broad, long-term issues and policies. (12, 26)

Suboptimal: Performance that is optimal according to the limited objectives or performance measures of a subunit of an organization, but that may be inconsistent with the goals of the organization as a whole. (132)

Subsystem: A component part of a system. (24, 241)

Summary report: A report that aggregates more detailed data within given categories, and only displays the sum. (111)

Supercomputer: A very fast and expensive computer, currently used mostly for scientific and engineering calculations but with growing use in large-scale optimization problems. (61)

Symbolic reference: The identification within a computer program of an operation or data element by a symbolic name (e.g., ADD or BALANCE) rather than machine-language code. (75)

Syntax: The set of rules governing the formation of statements in a language; for a computer language, the rules for naming variables, forming arithmetic or logical expressions, punctuation, and the like. (175)

System: An entity composed of interacting subunits that have a common purpose or set of global goals. (23, 241)

System design: The preparation of a detailed technical description of a system that provides a specified set of functional capabilities. (146)

System software: Computer programs that provide a common set of generic capabilities rather than application-specific functions. (67)

Systems concepts: The body of knowledge dealing with general characteristics and behavior of systems and how they can be made more efficient and effective. (240)

Tactical level: The middle level in an organization, which typically deals with medium-range decisions for a limited segment of the enterprise. (12, 24)

Telecommunications: The transmission of data over a network by means of an an electrical or optical channel. (5, 45)

Third-generation: The computer hardware and software in widespread use in the 1960s and 1970s, characterized by integrated microelectronic circuits and higher-level procedural languages. (76)

Third-party vendor: A vendor of software or add-on hardware for a computer produced by another manufacturer. (68, 162)

Timeliness: The provision of information having a recency and response time consistent with the needs of the user or the dynamic characteristics of the process being controlled. (209)

Time-sharing: A multiprogramming system that accommodates multiple users concurrently by allocating frequent small bursts of computing to each user, giving service equivalent to that of a dedicated small computer. (69)

Track: A narrow circular band on the surface of a storage disk on which data are recorded. (37)

Tradeoff: The extent to which performance with respect to a desired objective must be reduced in order to improve performance with respect to another objective. (146, 210)

Transaction: An event of significance to the MIS, such as the receipt of a customer order. (27, 85)

Transaction processing: The processing required to execute a transaction, such as updating the database and preparing outputs. (11)

Translator: *See* Language translator.

Turnkey system: A complete system ready for operation (requiring the user merely to "turn on the key"), including hardware, system software, and application programs. (69)

Unbundled pricing: A pricing policy in which a customer is charged separately for each product. *See* Bundled pricing. (68, 237)

Uncertainty: Lack of accurate knowledge about present or future events that will affect the outcome of a decision. (135)

Unit test: A test of a single program module performed before combining the module with other modules in a higher-level test. (149)

Updating: Changing the database to reflect all of the consequences of a transaction. (33)

User-friendly: The characteristic of a computer program that makes the system easy to learn or easy to use. (81, 175)

Utility program: A general-purpose program for performing a common data processing task, such as sorting a file into alphabetical sequence. (74)

Value chain: The sequence of steps taken by an organization in delivering a product or service from its suppliers through to its customers, with each step adding a portion of the final value of the end product or service. (295)

Value-added network (VAN): A communication service that leases raw bandwidth from a common carrier and adds additional services for its customers. (50)

Variable cost: A cost that changes relatively significantly with changes in the level of an activity. (233)

Variance: The difference between the total revenue received for services and the cost of providing the services at standard rates. (235)

VDU (Video display unit): *See* Cathode ray tube.

Verification: The process of comparing data that have been entered into a system with the original source document to determine the accuracy of the data. (29)

Voice mail: The storage of a recorded voice message in a form that can be retrieved by the recipient at his or her convenience (in a manner similar to electronic mail, but with voice output). (277)

Voice recognition: The automatic conversion of spoken words into meaningful digital form in order to enter data and instructions into an information system. (118)

Voice synthesis: The automatic conversion of words in digital form into spoken words (i.e., the inverse of voice recognition). (90)

Window: A partition of a physical display screen that provides an independent user interface for a task; with several windows on the same physical screen, a user can maintain concurrent dialogues with two or more separate programs (e.g., electronic mail and word processing) and communicate among them (e.g., incorporate an incoming message as part of a document). (81, 161)

Word processing: The use of a computer system to create and manipulate documents; the functions supported typically include text entry, editing, formatting, storage, retrieval, and printing. (63)

Word: A contiguous group of bits manipulated as a single entity by a computer, which may contain a group of binary-coded characters, a number represented in the binary number system, or a computer instruction. (65)

Workstation: A personal computer, often combining a powerful processor with a high-quality graphic display device. (47)

Write: To record data stored in the main memory of a computer on an external storage medium or onto a printed document. (36)

Subject Index

Note: Terms included in the Glossary are printed in boldface.